Paleofantasy

ALSO BY MARLENE ZUK

Sex on Six Legs

Riddled with Life

Sexual Selections

Paleofantasy

WHAT EVOLUTION REALLY TELLS US
ABOUT SEX, DIET, AND
HOW WE LIVE

Marlene Zuk

W. W. NORTON & COMPANY
New York • London

Printed in the United States of America
First Edition

For information about permission to reproduce selections from this book, write to Permissions, W. W. Norton & Company, Inc., 500 Fifth Avenue, New York, NY 10110

For information about special discounts for bulk purchases, please contact W. W. Norton Special Sales at specialsales@wwnorton.com or 800-233-4830

Manufacturing by Courier Westford
Book design by Chris Welch
Production manager: Devon Zahn

Library of Congress Cataloging-in-Publication Data

Zuk, M. (Marlene)
Paleofantasy : what evolution really tells us about sex, diet, and how we live / Marlene Zuk. — First edition.
pages cm
Includes bibliographical references and index.
ISBN 978-0-393-08137-4 (hardcover)
1. Human evolution. 2. Social evolution. 3. Hunting and gathering societies. 4. Prehistoric peoples—Food. 5. Prehistoric peoples—Sexual behavior. I. Title.
GN281.Z85 2013
599.93'8—dc23

2012047814

W. W. Norton & Company, Inc.
500 Fifth Avenue, New York, N.Y. 10110
www.wwnorton.com

W. W. Norton & Company Ltd.
Castle House, 75/76 Wells Street, London W1T 3QT

1 2 3 4 5 6 7 8 9 0

Contents

Paleofantasy

Introduction

The first thing you have to do to study 4,000-year-old DNA is take off your clothes. I am standing with Oddný Ósk Sverrisdóttir in the air lock room next to the ancient-DNA laboratory at Uppsala University in Sweden,[1] preparing to see how she and her colleagues examine the bones of human beings and the animals they domesticated thousands of years ago. These scientists are looking for signs of changes in the genes that allow us to consume dairy products past the age of weaning, when all other mammals lose the ability to digest lactose, the sugar present in milk. After that time, dairy products can cause stomach upsets. But in some groups of humans, particularly those from northern Europe and parts of Africa, lactase—the enzyme that breaks down lactose—lingers throughout life, allowing them to take advantage of a previously unusable food source. Oddný and her PhD supervisor, Anders Götherström, study how and when this development occurred, and how it is related to the domestication of cows for their meat and milk. They examine minute changes in genes obtained from radiocarbon-dated bones from archaeological sites around Europe.

The first step is to extract the DNA from the bones. But when examining genes from other humans, you must avoid contaminating the samples with your own genetic material. Suddenly I feel sullied by my own DNA and imagine it floating all around me, like infestive dust motes, needing to be contained as if it were the miasma of a terrible plague. Oddný, a tall, blonde Icelandic woman who looks like my image of a Valkyrie, at least if Valkyries are given to cigarette breaks and bouts of cheerful profanity, has brought a clean set of clothes for me to put on under the disposable space-suit-like outfit I need to wear in the lab. I have to remove everything except my underwear, including my jewelry. Götherström says it is the only time he ever takes off his wedding ring. I don the clean outfit, followed by the white papery suit, a face mask that includes a transparent plastic visor over my eyes, latex gloves, and a pair of slip-on rubber shoes from a pile kept in the neverland between the lab and the outside world. Anything else that goes into the lab—a flash drive for the computer, say—cannot go back in once it has gone out, to prevent secondary contamination of the facility. Finally, I put on a hairnet and tuck my hair underneath.

We enter the lab, where the first thing we do is stretch another pair of gloves over the ones we just put on. Oddný takes out a plastic bin of bone samples, each in its own zip-top bag. The bones themselves have been bleached and then irradiated with ultraviolet light to remove surface contamination. Before setting the bin on the counter, she wipes the surface with ethanol, followed by a weak bleach solution, and then with more ethanol. Apparently the saying that one can't be too careful is taken literally in this lab. "We all have to be kind of OCD to do this work," says Oddný, smiling. Or at least I think she is smiling under her mask.

To obtain the DNA, the bones are drilled and the powder from the interior is processed so that the genetic sequences inside are amplified—that is, replicated to yield a larger amount of material for easier analysis. Some bones are more likely to be fruitful than others; we heft the samples, since Oddný says that the most promising ones are heavy for their size, and glossy. Most of the samples are about 4,000 years old, but one of them is around 16,000 years old. It has already been rendered into powder, and I look at it closely, but it doesn't seem any different from the others. One of the pieces is a flat section of skull, while others are sections of leg or arm bones, or a bit of pelvis. Oddný and I wonder briefly who all these people were, and what their lives were like. The details of their experiences, of course, are lost forever. But the signature of what they were able to eat and drink, and how their diet differed from that of their—our—ancestors, is forever recorded in their DNA.

Other than simple curiosity about our ancestors, why do we care whether an adult from 4,000 years ago could drink milk without getting a stomachache? The answer is that these samples are revolutionizing our ideas about the speed at which our evolution has occurred, and this knowledge, in turn, has made us question the idea that we are stuck with ancient genes, and ancient bodies, in a modern environment. We can use this ancient DNA to show that we are not shackled by it.

The speed of evolution and our cave dweller past

Because we often think about evolution over the great sweep of time, in terms of minuscule changes over millions of years when we went from fin to scaly paw to opposable-thumbed hand, it is easy to assume that evolution always requires eons. That assumption in turn makes us feel that humans, who have gone from savanna to asphalt in a mere few thousand years, must be caught out by the pace of modern life, when we'd be much better suited to something more familiar in our history. We're fat and unfit, we have high blood pressure, and we suffer from ailments that we suspect our ancestors never worried about, like posttraumatic stress disorder and AIDS. Dr. Julie Holland, writing in *Glamour* magazine, counsels that if you "feel less than human," constantly stressed and rundown, you need to remember that "the way so many of us are living now goes against our nature. Biologically, we modern *Homo sapiens* are a lot like our cavewoman ancestors: We're animals. Primates, in fact. And we have many primal needs that get ignored. That's why the prescription for good health may be as simple as asking, What would a cavewoman do?"[2]

Along similar lines, here are some comments from readers of the *New York Times* health blog *Well*:

> Our bodies evolved over hundreds of thousands of years, and they're perfectly suited to the life we led for 99% of that time living in small hunting and gathering bands.[3]

> We are (like it or not) warm-blooded vertebrate mammals, i.e., part of the animal kingdom, and only in a very recent eyeblink of time become [*sic*] relatively free of the evolutionary pressures that shaped this species for millennia.[4]

> Probably goes all the way back to caveman days—women out gathering berries, sweeping up the place, generally always on the run. Cave Mr. Man out risking his neck, hunting a sabre tooth tiger or maybe a wooly mammoth, dragging the thing home, and then collapsing in a heap on the couch with a beer. I get it—makes sense.[5]

I am not suggesting that *Glamour* magazine or the readers of the *New York Times* have pinpointed the modern dilemma in its entirety. But it's hard to escape the recurring conviction that somewhere, somehow, things have gone wrong. In a time with unprecedented ability to transform the environment, to make deserts bloom and turn intercontinental travel into the work of a few hours, we are suffering from diseases our ancestors of a few thousand years ago, much less our prehuman selves, never knew: diabetes, hypertension, rheumatoid arthritis. Recent data from the Centers for Disease Control and Prevention (CDC) suggest that for the first time in history, the members of the current generation will not live as long as their parents, probably because obesity and associated maladies are curtailing the promise of modern medicine.

Some of this nostalgia for a simpler past is just the same old amnesia that every generation has about the good old days actually being all that good. The ancient Romans fretted about the young and their callous disregard for the hard-won wisdom of their elders. Several sixteenth- and seventeenth-century writers and philosophers famously idealized the Noble Savage, a being who lived in harmony with nature and did not destroy his surroundings. Now we worry about our kids as "digital natives," who grow up surrounded by electronics and can't settle their brains sufficiently to concentrate on walking the dog without simultaneously texting and listening to their iPods.

Another part of the feeling that the modern human is misplaced in urban society comes from the realization that people are still genetically close not only to the Romans and the seventeenth-century Europeans, but to Neandertals, to the ape ancestors Holland mentions, and to the small bands of early hominids that populated Africa hundreds of thousands of years ago. It is indeed during the blink of an eye, relatively speaking, that people settled down from nomadism to permanent settlements, developed agriculture, lived in towns and then cities, and acquired the ability to fly to the moon, create embryos in the lab, and store enormous amounts of information in a space the size of our handily opposable thumbs.

Given this whiplash-inducing rate of recent change, it's reasonable to conclude that we aren't suited to our modern lives, and that our health, our family lives, and perhaps our sanity would all be improved if we could live the way early humans did. Exactly what we mean by "the way early humans did" is a point of contention, and one I will return to in detail in Chapter 2, but the preconception—an erroneous one, as I will demonstrate—is the same: our bodies and minds evolved under a particular set of circumstances, and in changing those circumstances without allowing our bodies time to evolve in response, we have wreaked the havoc that is modern life.

In short, we have what the anthropologist Leslie Aiello, president of the renowned Wenner-Gren Foundation, called "paleofantasies."[6] She was referring to stories about human evolution based on limited fossil evidence, but the term applies just as well to the idea that our modern lives are out of touch with the way human beings evolved and that we need to redress the imbalance. Newspaper articles, morning TV, dozens of books, and self-help advocates promoting slow-food or no-cook diets, barefoot running, sleeping with our infants, and other measures large and small claim that it would be more natural, and healthier, to live more like our ancestors. A corollary to this notion is that we are good at things we had to do back in the Pleistocene, like keeping an eye out for cheaters in our small groups, and bad at things we didn't, like negotiating with people we can't see and have never met.

I am all for examining human health and behavior in an evolutionary context, and part of that context requires understanding the environment in which we evolved. At the same time, discoveries like those from Oddný's lab in Sweden and many more make it clear that we cannot assume that evolution has stopped for humans, or that it can take place only ploddingly, with tiny steps over hundreds of thousands of years. In just the last few years we have added the ability to function at high altitudes and resistance to malaria to the list of rapidly evolved human characteristics, and the stage is set for many more. We can even screen the entire genome, in great gulps of DNA at a time, looking for the signature of rapid selection in our genes.

To think of ourselves as misfits in our own time and of our own making flatly contradicts what we now understand about the way evolution works—namely, that rate matters. That evolution can be fast, slow, or in-between, and that understanding what makes the difference is far more enlightening, and exciting, than holding our flabby modern selves up against a vision—accurate or not—of our well-muscled and harmoniously adapted ancestors. The coming chapters will show just how much we know about that harmony, about the speed of evolution, and what these findings mean about the future of human evolution.

Our maladapted ancestors

The paleofantasy is a fantasy in part because it supposes that we humans, or at least our protohuman forebears, were at some point perfectly adapted to our environments. We apply this erroneous idea of evolution producing the ideal mesh between organism and surroundings to other life-forms too, not just to people. We seem to have a vague idea that long long ago, when organisms were emerging from the primordial slime, they were rough-hewn approximations of their eventual shape, like toys hastily carved from wood, or an artist's first rendition of a portrait, with holes where the eyes and mouth eventually will be. Then, the thinking goes, the animals were subject to the forces of nature. Those in the desert got better at resisting the sun, while those in the cold evolved fur or blubber or the ability to use fire. Once those traits had appeared and spread in the population, we had not a kind of sketch, but a fully realized organism, a fait accompli, with all of the lovely details executed, the anatomical t's crossed and i's dotted.

But of course that isn't true. Although we can admire a stick insect that seems to flawlessly imitate a leafy twig in every detail, down to the marks of faux bird droppings on its wings, or a sled dog with legs that can withstand subzero temperatures because of the exquisite heat exchange between its blood vessels, both are full of compromises, jury-rigged like all other organisms. The insect has to resist disease, as well as blend into its background; the dog must run and find food, as well as stay warm. The pigment used to form those dark specks on the insect is also useful in the insect immune system, and using it in one place means it can't be used in another. For the dog, having long legs for running can make it harder to keep the cold at bay, since more heat is lost from narrow limbs than from wider ones. These often conflicting needs mean automatic trade-offs in every system, so that each may be good enough but is rarely if ever perfect. Neither we nor any other species have ever been a seamless match with the environment. Instead, our adaptation is more like a broken zipper, with some teeth that align and others that gape apart. Except that it looks broken only to our unrealistically perfectionist eyes—eyes that themselves contain oddly looped vessels as a holdover from their past.

Even without these compromises from natural selection acting on our current selves, we have trade-offs and "good enough" solutions that linger from our evolutionary history. Humans are built on a vertebrate plan that carries with it oddities that make sense if you are a fish, but not a terrestrial biped. The paleontologist Neal Shubin points out that our inner fish constrains the human body's performance and health because adaptations that arose in one environment bedevil us in another.[7] Hiccups, hernias, and hemorrhoids are all caused by an imperfect transfer of anatomical technology from our fish ancestors. These problems haven't disappeared for a number of reasons: just by chance, no genetic variants have been born that lacked the detrimental traits, or, more likely, altering one's esophagus to prevent hiccups would entail unacceptable changes in another part of the anatomy. If something works well enough for the moment, at least long enough for its bearer to reproduce, that's enough for evolution.

We can acknowledge that evolution is continuous, but still it seems hard to comprehend that this means each generation can differ infinitesimally from the one before, without a cosmic moment when a frog or a monkey looked down at itself, pronounced itself satisfied, and said, "Voilà, I am done." Our bodies therefore reflect a continuously jury-rigged system with echoes of fish, of fruit fly, of lizard and mouse. Wanting to be more like our ancestors just means wanting more of the same set of compromises.

When was that utopia again?

Recognizing the continuity of evolution also makes clear the futility of selecting any particular time period for human harmony. Why would we be any more likely to feel out of sync than those who came before us? Did we really spend hundreds of thousands of years in stasis, perfectly adapted to our environments? When during the past did we attain this adaptation, and how did we know when to stop?

If they had known about evolution, would our cave-dwelling forebears have felt nostalgia for the days before they were bipedal, when life was good and the trees were a comfort zone? Scavenging prey from more formidable predators, similar to what modern hyenas do, is thought to have preceded, or at least accompanied, actual hunting in human history. Were, then, those early hunter-gatherers convinced that swiping a gazelle from the lion that caught it was superior to that newfangled business of running it down yourself? And why stop there? Why not long to be aquatic, since life arose in the sea? In some ways, our lungs are still ill suited to breathing air. For that matter, it might be nice to be unicellular: after all, cancer arises because our differentiated tissues run amok. Single cells don't get cancer.

Even assuming we could agree on a time to hark back to, there is the sticky issue of exactly what such an ancestral nirvana was like. Do we follow the example of the modern hunter-gatherers living a subsistence existence in a few remaining parts of the world? What about the great apes, the animals that most closely resemble the ancestors we (and they) split off from millions of years ago? How much can we deduce from fossils? People were what anthropologists call "anatomically modern," meaning that they looked more or less like us, by about 200,000 years ago, but it's far less clear when "behaviorally modern" humans arose, or what exactly they did. So, trying to deduce the classic lifestyle from which we've now deviated is itself a bit of a gamble. In his book *Before the Dawn*, science writer Nicholas Wade points out, "It is tempting to suppose that our ancestors were just like us except where there is evidence to the contrary. This is a hazardous assumption."[8]

You might argue that hunter-gatherers, or the cavemen of our paleofantasies, were better adapted to their environment simply because they spent many thousands of years in it—much longer than we've spent sitting in front of a computer or eating Mars bars. That's true for some attributes, but not all. Continued selection in a stable environment, as might occur in the deep sea, can indeed cause ever more finely honed adaptations, as the same kinds of less successful individuals are weeded out of the population. But such rock-solid stability is rare in the world; the Pleistocene varied considerably in its climate over the course of thousands of years, and when people move around, even small shifts in the habitat in which they live, going from warm to cool, from savanna to forest, can pose substantially new evolutionary challenges. Even in perfectly stable environments, trade-offs persist; you can't give birth to large-brained infants and also walk on two legs trouble-free, no matter how long you try.

Incidentally, it's important to dispel the myth that modern humans are operating in a completely new environment because we only recently began to live as long as we do now, whereas our ancestors, or the average hunter-gatherer, lived only until thirty or forty, and hence never had to experience age-related diseases.

While it is absolutely true that the average life span of a human being has increased enormously over just the last few centuries, this does not mean that thousands of years ago people were hale and hearty until thirty-five and then suddenly dropped dead.

An average life expectancy is just that—an average of all the ages that the people in the population attain before they die. A life expectancy of less than forty can occur without a single individual dying at or even near that age if, for example, childhood mortality from diseases such as measles or malaria is high—a common pattern in developing countries. Suppose you have a village of 100 people. If half of them die at age five, perhaps from such childhood ailments, twenty die at age sixty, and the remaining thirty die at seventy-five, the average life span in the society is thirty-seven, but not a single person actually reached the age of thirty hale and hearty and then suddenly began to senesce. The same pattern writ large is what makes the life expectancy in developing countries so shockingly low. It isn't that people in sub-Saharan Africa or ancient Rome never experienced old age; it's that few of them survived their childhood diseases. Average life expectancy is not the same thing as the age at which most people die. Old age is not a recent invention, but its commonness is.

The pace of change

If we do not look to a mythical past utopia for clues to a way forward, what next? The answer is that we start asking different questions. Instead of bemoaning our unsuitability to modern life, we can wonder why some traits evolve quickly and some slowly. How do we know what we do about the rate at which evolution occurs? If lactose tolerance can become established in a population over just a handful of generations, what about an ability to digest and thrive on refined grains, the bugaboo of the paleo diet? Breakthroughs in genomics (the study of the entire set of genes in an organism) and other genetic technologies now allow us to determine how quickly individual genes and gene blocks have been altered in response to natural selection. Evidence is mounting that numerous human genes have changed over just the last few thousand years—a blink of an eye, evolutionarily speaking—while others are the same as they have been for millions of years, relatively unchanged from the form we share with ancestors as distant as worms and yeast. The pages to come will explore which genes and traits have changed, which have not, how we know, and why it matters.

What's more, a new field called experimental evolution is showing us that sometimes evolution occurs before our eyes, with rapid adaptations happening in 100, 50, or even a dozen or fewer generations. Depending on the life span of the organism, that could mean less than a year, or perhaps a quarter century. It is most easily demonstrated in the laboratory, but increasingly, now that we know what to look for, we are seeing it in the wild. And although humans are evolving all the time, it is often easier to see the process in other kinds of organisms. Humans are not the only species whose environment has changed dramatically over the last few hundred years, or even the last few decades. Some of the work my students and I have been doing on crickets found in the Hawaiian Islands and in the rest of the Pacific shows that a completely new trait, a wing mutation that renders males silent, spread in just five years, fewer than twenty generations.[9] It is the equivalent of humans becoming involuntarily mute during the time between the publication of the Gutenberg Bible and *On the Origin of Species*. This and similar research on animals is shedding light on which traits are likely to evolve quickly and under what circumstances, because we can test our ideas in real time under controlled conditions.

Over the last decade, our understanding of such rapid evolution, also called "evolution in ecological timescales," has increased enormously. And studying the rate of evolution also has practical implications. For example, fishermen often take the largest specimens of salmon or trout from streams and rivers. Fish usually need to reach a certain size before becoming sexually mature and capable of reproduction, after which growth slows down. Like other animals, fish show a trade-off between large size and time of reproduction: if you wait to be large before producing offspring, you probably will be able to produce more of them, and having greater numbers of offspring is favored by evolution, but you also risk dying before you are able to reproduce at all. But where overfishing has removed a substantial portion of a population, the average size of fish is now substantially smaller, because the fishermen have inadvertently selected for earlier reproduction, and evolution has favored fish that get to the business of sex sooner. It's not just that the larger fish have all been taken; it's that the fish are not reaching such sizes to begin with. The genes responsible for regulating growth and size at sexual maturity are now different because evolution has occurred. To bring back the jaw-dropping trophy fish of decades past, scientists say, people will have to change their ways.

It's common for people talk about how we were "meant" to be, in areas ranging from diet to exercise to sex and family. Yet these notions are often flawed, making us unnecessarily wary of new foods and, in the long run, new ideas. I would not dream of denying the evolutionary heritage present in our bodies—and our minds. And it is clear that a life of sloth with a diet of junk food isn't doing us any favors. But to assume that we evolved until we reached a particular point and now are unlikely to change for the rest of history, or to view ourselves as relics hampered by a self-inflicted mismatch between our environment and our genes, is to miss out on some of the most exciting new developments in evolutionary biology.

The influential twentieth-century biologist George Gaylord Simpson wrote a book called *Tempo and Mode in Evolution*, published in 1944. It is admirable from many perspectives, not least of which is the distinction it makes between the rate at which evolution occurs (tempo) and the pattern of evolution itself (mode). Simpson, a paleontologist by training, saw the work as an attempt to merge the then-new field of genetics with his own—a procedure he admitted to be "surprising and possibly hazardous":

> Not long ago paleontologists felt that a geneticist was a person who shut himself in a room, pulled down the shades, watched small flies disporting themselves in milk bottles, and thought that he was studying nature. A pursuit so removed from the realities of life, they said, had no significance for the true biologist. On the other hand, the geneticists said that paleontology had no further contributions to make to biology, that its only point had been the completed demonstration of the truth of evolution, and that it was a subject too purely descriptive to merit the name "science." The paleontologist, they believed, is like a man who undertakes to study the principles of the internal combustion engine by standing on a street corner and watching the motor cars whiz by.[10]

We still sometimes think that paleontology, or evolution at grand scales—the rise and fall of dinosaurs, the origin of land animals—has little in common with the minuscule goings-on of the genes from one generation to the next. But Simpson recognized that the two processes, while having some distinctive components, are still linked, and that the disporting flies exhibit many of the same characteristics as those million-year-old bones. It's just that the scale of measurement differs.

The title of Simpson's book is particularly germane to my argument here, calling up as it does a rather orchestral view of evolution, with allegro and adagio components. Seeing evolution without appreciating its variously fast and slow parts is like making all the movements of a symphony happen at the same pace; you get the same notes, but most of the joy and subtlety are missing. New advances in biotechnology have made the merger of paleontology and genetics more feasible than Simpson could have imagined. We have not yet cloned dinosaurs à la *Jurassic Park*, but we are not too far off. And this merger means that we can examine not only what evolution has wrought, but also the pace at which it operates, just as Simpson hoped.

Change doesn't always do you good

At the same time that we wistfully hold to our paleofantasy of a world where we were in sync with our environment, we are proud of ourselves for being so different from our apelike ancestors. Animals like crocodiles and sharks are often referred to as "living fossils" because their appearance is eerily similar to that of their ancestors from millions of years earlier that are preserved in stone. But there is sometimes a tone of disparagement in the term; it is as though we pity them for not keeping up with trends, as if they are embarrassing us by walking (or swimming) around in the evolutionary equivalent of mullet haircuts and suspenders. Evolving more recently, so that no one would mistake a human for our predecessors of even a couple of million years back, seems like a virtue, as if we improved ourselves while other organisms stuck with the same old styles their parents wore.

Regardless of the shaky ground on which that impression lies, we don't even win the prize for most recent evolution; in fact, we lose by a wide margin. Strictly speaking, according to the textbook definition of evolution as a change in gene frequencies in a population, many of the most rapidly evolving species, and hence those with the most recent changes, are not primates but pathogens, the disease-causing organisms like viruses and bacteria. Because of their rapid generation times, viruses can produce offspring in short order, which means that viral gene frequencies can become altered in a fraction of the time it would take to do the same thing in a population of humans, zebras, or any other vertebrate.

Evolution being what it is—namely, without any purpose or intent—evolving quickly is not necessarily a good thing. Often the impetus behind rapid evolution in nonhuman organisms is a strong and novel selective agent: a crop is sprayed with a new insecticide, or a new disease is introduced to a population by a few individuals who stray into its boundaries. Those who are resistant, sometimes an extremely tiny minority, survive and reproduce, while the others perish. These events are not confined to crops, or even to nonhumans. Some estimates of death rates from the medieval outbreak of bubonic plague called the Black Death in Europe have gone as high as 95 percent.

Natural selection thus produces a bottleneck, through which only the individuals with the genes necessary for survival can pass. The problem is, that bottle also squeezes out a lot of other genetic variation along with the genes for susceptibility to the insecticide or the ailment. Suppose that genes for eye color or heat tolerance or musical ability happen to be located near the susceptibility genes on the chromosome. During the production of sex cells, as the chromosomes line up and the sperm and egg cells each get their share of reshuffled genes, those other genes will end up being disproportionately likely to be swept away when their bearer is struck down early in life by the selective force—the poison or pathogen. The net result is a winnowing out of genetic variants overall, not just those that are detrimental in the face of the current selection regime.

Future evolution, and the downside to immortality

Evolution is constantly at work, altering a gene here or a set of co-occurring attributes there. It's not always visible, at least not at first, but it's still happening. And it provides a little-considered flaw to that long-sought goal of humanity: immortality. Imagine that you, like a character in one of those vampire novels that are so popular these days, can live forever. Day in and day out, as the seasons change and the years go by, you remain deathless and unaltered, while those around you wither and die. Except for the inconvenience of having to hide your ageless physique from the mortals around you, and the necessity of catching up with new fashions not once during a single period of youth but over and over again as hemlines rise and fall, it would be perfect, right?

Or maybe not. And not for the usual literary-device reasons, like losing your purpose in life, lacking a need to leave your mark before you expire, or having to watch loved ones succumb to the ravages of time. No. As generations came and went, it would become increasingly apparent that the problem was the inability to evolve. Individuals never can evolve, of course; members of a population just leave more or fewer genetic representations of themselves. But since we are never around to see more than a couple of generations before or after us, we don't notice the minute changes that are occurring in the rest of the group. After a while, and not all that long a while at that, your fifteenth-century vampire self would start looking, well, maybe not like a Neandertal, but just a bit different. You would be shorter than your peers, for example. And even if you didn't look different, your insides would lack those latest-model advances, those features that make the new version the one to buy, such as resistance to newfangled diseases like malaria. Natural selection happened while you just kept on being a bloodsucker.

Evolutionary biologist Jerry Coyne, author of *Why Evolution Is True* and an eponymous website, says that the one question he always gets from public audiences is whether the human race is still evolving.[11] On the one hand, modern medical care and birth control have altered the way in which genes are passed on to succeeding generations; most of us recognize that we wouldn't stand a chance against a rampaging saber-toothed tiger without our running shoes, contact lenses, GPS, and childhood vaccinations. Natural selection seems to have taken a pretty big detour when it comes to humans, even if it hasn't completely hit the wall. At the same time, new diseases like AIDS impose new selection on our genomes, by favoring those who happen to be born with resistance to the virus and striking down those who are more susceptible.

Steve Jones, University College London geneticist and author of several popular books, has argued for years that human evolution has been "repealed" because our technology allows us to avoid many natural dangers.[12] But many anthropologists believe instead that the documented changes over the last 5,000–10,000 years in some traits, such as the frequency of blue eyes, means that we are still evolving in ways large and small. Blue eyes were virtually unknown as little as 6,000–10,000 years ago, when they apparently arose through one of those random genetic changes that pop up in our chromosomes. Now, of course, they are common—an example of only one such recently evolved characteristic. Gregory Cochran and Henry Harpending even suggest that human evolution as a whole has, on the contrary, accelerated over the last several thousand years, and they also believe that relatively isolated groups of people, such as Africans and North Americans, are subject to differing selection.[13] That leads to the somewhat uncomfortable suggestion that such groups might be evolving in different directions—a controversial notion to say the least.

The "fish out of water" theme is common in TV and movies: city slickers go to the ranch, Crocodile Dundee turns up in Manhattan, witches try to live like suburban housewives. Misunderstandings and hilarity ensue, and eventually the misfits either go back where they belong or learn that they are not so different from everyone else after all. Watching people flounder in unfamiliar surroundings seems to be endlessly entertaining. But in a larger sense we all sometimes feel like fish out of water, out of sync with the environment we were meant to live in. The question is, did that environment ever exist?

1

Cavemen in Condos

In 2010 the *New York Times* ran an article titled "The New Age Cavemen and the City," about modern-day followers of a supposedly evolution-based lifestyle.[1] These people, mainly men, subsist largely on meat (they apparently differ about whether or not it is best, or even acceptable, to cook it), eschew any foods requiring that newfangled practice of cultivation, and exercise in bursts of activity intended to mimic a sprint after escaping prey. The *Sydney Morning Herald* ran a similar article, with one adherent noting, "The theory is that you only eat what our ancestors ate 10,000 years ago. It's what you could get with a stick in the forest."[2] Frequent blood donation is also practiced, stemming from the idea that cavemen were often wounded and hence blood loss would have been common. (University of Wisconsin anthropologist John Hawks noted that the resulting photograph of three practitioners of this paleo way of life looks "like the cast of Pleistocene *Twilight*."[3]) Surprisingly, New York City turns out to be a hospitable place to practice such principles, partly because it is easier to walk to most of one's destinations. One of those profiled does his walking "dressed in a tweed coat and Italian loafers,"[4] but the lack of adherence to an ancestral wardrobe of (presumably) skins and hides goes unremarked.

The "paleo" (a word many prefer to "caveman") practitioners are a varied bunch; in addition to the group described in the *New York Times*, numerous diet books and blogs, exercise programs, and advice columns exist to help those trying not just to be healthy, but to do so in a way that hews to how our early human ancestors would have lived. According to a commenter on one such site, Cavemanforum.com, "I see more and more mistakes in moving away from paleo life. All these things we need, to feel happy, to be healthy—it sounds stupid, but I've starting to feel like agriculture really was the biggest mistake we ever did. Of course we can't bring the times back, but in a strange way I wish we could. The solutions to our problems lay there. Not just food. I feel like we messed up, and we are paying for it."[5]

References to masculinity are common (one blog links to "The art of manliness"), but Paleochix.com does offer a more female-oriented approach, with articles on skin care and motherhood. A Google search for "caveman lifestyle" garners over 200,000 hits. In *The Omega Diet*, Artemis P. Simopoulos and Jo Robinson claim, "Human-like creatures have existed on this planet for as long as four million years, and for roughly 99 percent of this time, they were hunters and gatherers . . . This means that when we're sitting down to lunch, our stone-age bodies 'expect' to be fed the same types and ratios of fat that nourished our cave-dwelling ancestors."[6] And on Diabetescure101.com we see this statement: "When you realize you are stuck with a caveman body and realize you have been doing all sorts of 'Modern Man' things to it to screw up the system, and [as] a result it is not working, then you will take the time to stop and figure out what needs to be changed."[7]

The articles were clearly a bit tongue in cheek, and even the most ardent followers of the paleo life are not seriously trying to live exactly like people would have 10,000 years ago or more. For one thing, where are all those caves going to come from? And does it count if the blood loss occurs via needle rather than bear tooth?

At least one commenter on the *New York Times Well* blog sounds a skeptical note: "It is idiotic to model one's behavior on the practices of pre-modern humans on the belief that it will make you live longer, or result in improved quality of life in later years. You are not a pre-modern human. Get over it."[8]

Anthropologist Greg Downey of Macquarie University in Australia once mused to me that it's interesting that people never yearn for the houses of yesteryear, or yester-epoch; we seem to have abandoned mud huts quite happily.[9] But the proliferation of essays and conversations does show the appeal of trying to take on at least a few attributes of our ancestors, and at least some people are quite convinced that modern civilization has led us astray. The catch is, if we want to go back to a healthier way of life, what exactly should we emulate? How did those ancestors live anyway, and where do we get our ideas about early humans?

Furthermore, even if we reject the idea of living like a caveman, is it still reasonable to assume that we have spent the last 10,000 years or so being whipsawed through history without sufficient time to adapt? Asking questions about what it would be like if we lived more like our Paleolithic ancestors can lead to much more interesting questions than whether or not we should be eating grains or wearing glasses (several of the paleo proponents are very disparaging about the use of such devices[10]). Really understanding our history means understanding which human traits arose when, and which ones are likely to have changed recently. It also means deciding what we can use as a model for our own evolution, given that our evidence is so limited. Bones fossilize, but few of those have survived, and our ancestral behaviors, including our social arrangements, our love lives, and the way we raised our children, leave no physical traces. Increasingly, however, we can reconstruct our history by examining our genes—an undertaking that reveals a much more complex story than the assumption that our DNA has remained unchanged since the advent of agriculture. It all comes down to the pace of evolution.

On evolution and pinnacles

We're all familiar with those cartoons that show evolutionary milestones, with a fish morphing into a lizard that crawls onto the land, followed by various types of mammals, and always concluding with a human, often clutching a spear, gazing off into the distance. Some versions start with humans at the outset, and in a graphic depiction of paleofantasy, show a knuckle-walking ape transforming into a beetle-browed guy with a club, followed by a rather well-muscled figure in a loincloth changing into a slouching paunchy fellow bent over a computer. Even taken as caricatures, these images contain a lot to object to: For one thing, they virtually never show women unless they are trying to make a point about sex roles, as if men did all the important evolving for our species. For another, they assume that evolution proceeded in a straight line, with each form leading seamlessly and inevitably into the next. But the real problem is that these evolutionary progressions make it seem as if evolution, well, progresses. And that simply isn't the case.

It is true that natural selection weeds out the individuals, and genes, less suited to the environment, leaving behind the ones that can manage to survive and reproduce. And what we see now—halibut with eyes that migrate during development so that both end up on the same side of the fish as it lies on the sandy ocean floor, moths with tongues exactly the size of the tubular flowers they feed on—looks awfully perfect, as if the fishy or mothlike progenitors were intending to go there all the time. But in fact, as I mentioned in the Introduction, organisms are full of trade-offs, and the solutions we see are only single possibilities among many. Halibut ancestors weren't striving to be flat, early forms of moths weren't all about sipping nectar from only the longest flowers, and Miocene apes weren't trying to end up driving cars and paying taxes, or even coming down out of the trees. Other paths were possible and might have led to flatfish with even better camouflage, or people with teeth that were not so prone to decay. So, that phrase "we evolved to . . ."—eat meat, have multiple sexual partners, or even just walk upright—is misleading, at least if it's taken to mean that we were *intending* to get there.

What's more, evolution can and often does occur without natural selection. Species change just because the precise combinations of genes in populations can shift, as individuals move from place to place or small accidents of fate wipe out whole sets of characteristics that are then no longer available for selection to act on. All of this means that we aren't at a pinnacle of evolution, and neither are the moths or halibut. And those cartoons, amusing though they can be, are simply wrong.

How we got here: A (very) brief history of human evolution

Before we can look at how fast humans are changing, we need to see where we came from. Although anthropologists still argue about many of the details of human evolution, they largely agree that more-humanlike primates arose from ape ancestors about 6 million years ago, when the Earth was cooling. We did not, of course, arise from apes that are identical, or even all that similar, to the modern chimpanzees, gorillas, and orangutans we see today, but from a different form entirely. Because of the sparseness of the fossil record, it is not clear exactly which kind of ape the earliest humans came from, but the first hominins (a name for the group that includes humans and our ancestors, but not chimpanzees, bonobos, orangutans, or gorillas) were different from their ape ancestors. They walked on two legs, rather than on all fours, and this bipedal mode of perambulating led to a variety of other skeletal changes to which we are arguably still adjusting, as evidenced by problems like lower

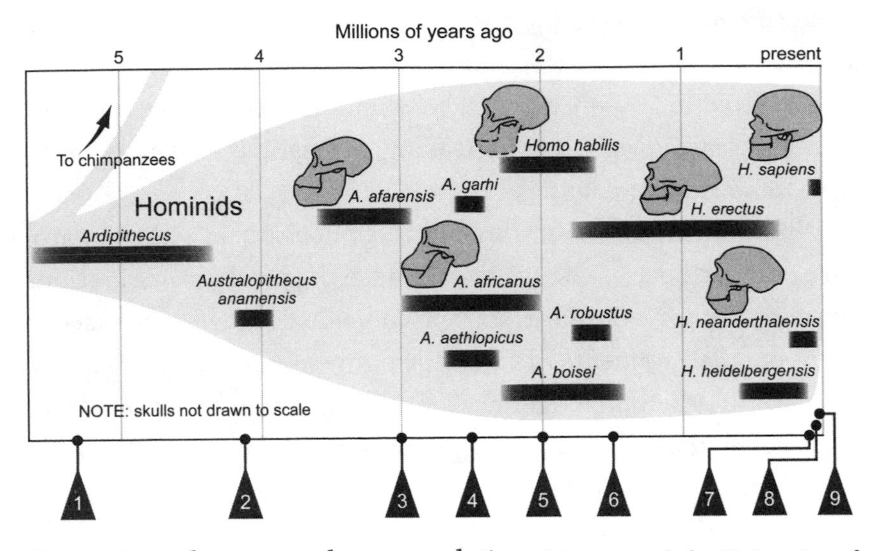

The major milestones in human evolution. (Courtesy of the University of California Museum of Paleontology—Understanding Evolution, http://evolution.berkeley.edu)

back pain. And they ate different kinds of foods as they moved into the woods and savannas instead of the forests of more tropical Africa, leading to changes in the teeth and skull as natural selection acted on each generation.

Between 4 and 2 million years ago, several different hominin species coexisted in Africa, including some in the genus *Homo*, which is the same genus to which contemporary humans belong. Modern humans have a number of characteristics that distinguish them from modern apes, including a relatively large brain, that bipedal walking mode, a long period of time spent as a juvenile dependent on others, and symbolic language that helps us communicate about a complex culture that includes material objects such as tools. Other animals share some of these traits with us—cetaceans, for example, also have relatively large brains for their body size—but taken as a whole, the list demarcates humans as humans.

Some anthropologists like to select one or more of the characteristics as key, such as the use of fire for cooking food, which Harvard anthropologist Richard Wrangham believes led to a cascade of other evolutionary events that were essential to our becoming modern humans.[11] Others focus on selection for large brains and longer child development. Bipedalism and the ability to walk upright are also important features of hominins, but while it certainly seems as if having one's hands free allows all kinds of other human attributes, like carrying tools or weapons around and hugging one's neighbors, these changes don't seem to have been the critical trigger for the other traits, like larger brains, since ancestral apes were bipedal long before the skulls of early hominins showed signs of increased brain size.

Regardless of your favorite demarcating trait, exactly when any one of these characteristics arose is still hotly contested among archaeologists, and the goalposts are shifting all the time. Until 2010, tool use, one of those key signs of technological advancement in human evolution, was thought to have arisen less than 3 million years ago. But careful examination of two animal bones from 800,000 years earlier, at the time when the species made famous by the skeleton called Lucy, *Australopithecus afarensis*, was roaming Africa, now suggests that ancestral humans might have been using stone tools to butcher meat as long as 3.4 million years ago. The rib and thigh bones of those animals carry tiny marks that sophisticated imaging techniques indicate were made by stone, not the teeth of predatory animals simply munching on their prey.[12] Lucy and her contemporaries, however, were relatively small-brained compared with the various species of *Homo*, our own genus, that came later, which presumably means that carving up the meat one has hunted doesn't necessarily require a lot of intelligence.

Other anthropologists are not convinced by the admittedly scanty fossil evidence, and another study, published later in 2010, makes the case that the marks could have been produced by decay, the hooves of animals stepping on the bones, or other innocuous sources.[13] More information is clearly needed before the jury is in on when humans first started to use tools. Nevertheless, estimates for the first use of fire have similarly been pushed back in time, along with the first use of ground grains for flour. In addition to being important and interesting from the standpoint of reconstructing our family history, the shifting benchmarks underscore the riskiness of declaring how long humans have spent doing any one thing, whether that thing is hunting, using tools, or cultivating food. In turn, this uncertainty means that we are on shaky ground declaring that humans have been doing certain things, such as eating grains, for "only" short periods, and hence cannot be well adapted to the new environment. How long is long enough, and how short is too short?

To understand how far we have come, we need to examine what life was like for the early hominins. As our ancestors moved to the savannas and woods of Africa, the climate was becoming more seasonal, with rain that was more likely to fall only at certain times of year. This seasonality meant that foods available during dry periods, like underground tubers and meat, became more important. The foods eaten, in turn, dictate the methods used to obtain them, and when it comes to humans, the big question is when and how people started hunting. Anthropologists used to think that early hominins did not eat much meat, because hunting is such a sophisticated activity (at least the way it's practiced by people today) that our smaller-brained early ancestors were not thought to be capable of it. But over the last couple of decades, chimpanzees, which have smaller brains than *Australopithecus afarensis*, have been found to hunt regularly, preying on monkeys, antelope, and wild pigs, suggesting that you don't have to be such a brainiac to be a predatory primate.

Such prey animals are of reasonable size—big enough, in fact, to share. And food sharing is also a significant milestone in the evolution of human social behavior. As anthropologists Rob Boyd and Joan Silk at Arizona State University note, "hunting makes sharing necessary, and sharing makes hunting feasible."[14] What they mean is that being able to bring down a deer is about more than venison. Hunting is a chancy business, with most predators, including wolves and lions, scoring a lot more misses than hits. But if individual hunters share what they kill, and if other people in the group hand out the more steady source of food provided by vegetable matter, the risk of starving to death after a run of bad luck in hunting is reduced. So if early humans hunted, they probably had the complex social interactions that go along with dividing up the kill. Food sharing might also have been linked to divisions of labor within a group, although whether the men necessarily did all the hunting while the women stayed home and dug for roots or picked berries is not at all clear, as I will discuss in Chapter 7.

Neither hunting large game nor finding wild foods like fruits, tubers, or honey is easy, and the skills involved can take many years to master. Once such difficult ways of getting food became commonplace among early humans, a feedback loop was possible: natural selection favored distinctive traits, like large brains and a longer time spent as a juvenile, because those in turn made the foraging easier, which in turn meant selection for ever more efficient ways to get food. By 1.8 million years ago, at the beginning of the Pleistocene, more modern hominins were established in Africa, though the earliest *Homo sapiens*, our own species, did not appear until a mere 100,000 years ago.

One of these early members of our genus, *Homo ergaster*, is noteworthy because it may have marked the appearance of the first long-distance runner in our lineage. Although earlier hominins were also bipedal at least some of the time, *Homo ergaster* was tall (one skeleton with the fetching name of KNM-WT 15000 was 5 feet 4 inches tall as a young boy, projected to stand at 6 feet when he was fully grown) and had the long legs, narrow hips, and more barrel-shaped chest of modern running humans, as opposed to the long arms and stubbier legs of earlier forms that probably still spent a considerable amount of time in the trees.

The running is significant not just because it contributes to the controversy over "natural" forms of exercise that I will discuss in Chapter 6, but because some anthropologists believe it was linked to the evolution of larger brains. Running would allow ancestral humans to get to carcasses left by predators like lions before other scavengers such as hyenas, and hence gain access to high-quality protein. Such concentrated food can fuel the demands of large brains, which are physiologically expensive to maintain. To obtain and process game animals, *Homo ergaster* made more sophisticated tools than earlier humans had made, including hand axes with nifty double-faced blades and sharp, narrow points. Interestingly, *Homo ergaster* also shared a doubtless unwelcome advancement with modern humans: intestinal parasites. Recent studies of the human tapeworm, a gut inhabitant that is specialized on humans and the animals they eat, show that at least two kinds of tapeworms date back to 1.7 million years ago, long before animal domestication.[15]

Homo ergaster spread from Africa into Europe, with other species of *Homo* also appearing in various parts of Eurasia. Between 900,000 and 130,000 years ago, during the Pleistocene, the Earth became cooler, with long periods when glaciers covered large parts of the world. While the glaciers covered some areas, deserts spread over northern Africa, separating the populations of plants and animals, including humans. Anthropologists do not agree about exactly how to classify the fossil humans found from this period, but it is clear that brain size evolved substantially from that of *Homo ergaster*; and one kind of early human, *Homo heidelbergensis*, hunted and butchered large animals such as mammoths. By 300,000 years ago, humans had produced even more finely crafted stone tools, including some that were suitable for attaching to handles—a marked improvement that allows much greater force to be applied to the tool.

The Neandertals, reconstructions of whom are probably responsible for most people's conceptions of what cavemen supposedly looked like, lived in Europe and eastern Asia from 127,000 until about 30,000 years ago. They were not human ancestors, meaning that their lineage did not lead directly to that of modern humans, but because they lived in Europe and many paleontologists found their own European homeland a more convenient place to dig for fossils than many parts of Africa and Asia, we know more about them than we do about any other early human.

The recent discovery that we humans share a small proportion of our genes with the Neandertals,[16] suggesting that early *Homo sapiens* and Neandertals mated with each other, has renewed interest in the lives of these hominins. The fossils reveal skulls with big front teeth that show heavy wear, perhaps because the Neandertals used to pull their meat through their incisors. The Neandertals also had rather stocky frames; according to Boyd and Silk, "A comparison with data on Olympic athletes suggests that Neanderthals most closely resembled the hammer, javelin, and discus throwers and shot-putters."[17] Such a build is in keeping with a species that lives in cold climates; shorter limbs help conserve body heat.

Neandertals had large brains—larger in an absolute sense, not necessarily relative to body size, than those of modern humans—and made complex stone tools that were probably used to hunt elk, deer, bison, and other large game animals. I say "probably" because, as with all conclusions drawn about early humans, this one is indirect. We find stone tools, and we find the bones of animals with marks on them from those tools. Did the Neandertals kill the animals, or could they have scavenged them from other predators? The consensus is that they were indeed killing the prey, but the data are not conclusive. Hunting or scavenging aside, the Neandertal life was not an idyllic one. Most Neandertals died before the age of fifty, and many of the adult skeletons show signs of diseases such as arthritis, gum disease, and deformed limbs.

Anatomically modern humans, which first appeared on the scene about 100,000 years ago, not only looked different from the Neandertals; they had a more complex social life, using materials that came from far away to make their tools, which suggests that they traded with distant people. Unlike the Neandertals, they built shelters, and they created art and buried their dead. They certainly hunted game, but how much they relied on meat as opposed to plant foods is unclear, and almost certainly it varied depending on what part of the world they lived in—humans migrated from Africa throughout much of the rest of the world during the Paleolithic—and what time period is being considered. Bones and tools preserve much better than stems, seeds, or pits, but that does not mean that early humans relied more on meat than on other food sources.

About 50,000 years ago, the fossil tools and other cultural accoutrements found in Europe underwent a marked change. Although people had looked like, well, people, for a long time, humans began to make more elaborate tools, clothing, and shelters, and at some point they began to use the symbolic communication that we call language. When and how language arose is, again, a hotly debated topic. Linguists are pessimistic that we will ever be able to trace the origin of modern language back further than 5,000 or at most 10,000 years, though it's very likely that humans were using some form of symbolic spoken communication well before then.

Indeed, the recent discovery of a gene called *FOXP2*, which is present in virtually identical form in all humans, has led scientists to believe they may have uncovered an essential component in the evolution of language. *FOXP2* occurs in many animals, including mice, and chimpanzees and gorillas possess identical forms of the gene. Defects in human *FOXP2* are associated with an inability to use words correctly. Biologist Svante Pääbo and colleagues at the Max Planck Institute for Evolutionary Anthropology in Leipzig, Germany, examined the small differences within the gene in chimps and gorillas, as well as in humans from several parts of the world, and discovered that *FOXP2* changed rapidly after humans and chimpanzees split off from our common ancestor, perhaps because strong selection made the ability to use language advantageous. Pääbo then calculated that the current version of the gene appeared in the human lineage within the last 200,000 years.[18]

Although other species of *Homo* were living in Europe and Asia over 100,000 years ago, modern humans, *Homo sapiens*, moved out of Africa to populate the rest of the world only about 60,000 years ago. All of the people on Earth today are descended from a rather small number of Africans. This limited population of origin is probably why humans are much less variable genetically than are many other species, including our close relatives the chimpanzees. We tend to think we are terribly different from each other while every chimp in the zoo looks alike, but our genes tell a different story. Pluck any two people at random, even from a relatively large population like that of southern Europe, and sequence their DNA, and you will find that their genes differ less than the genes of two chimpanzees from central Africa.

Bushmen, bones, and "chimpiness"

The next big step in human evolution after the great expansion from Africa was the beginning of agriculture, which facilitated larger populations and the eventual establishment of towns, social classes, and other modern accoutrements. But the time before that transition, from perhaps 60,000 to 10,000 years ago, when humans were living as hunters and foragers, is what inspires our paleofantasies. This is the time that some evolutionary psychologists call the Environment of Evolutionary Adaptedness, or EEA, when humans became what we are today.[19] In Chapter 2 I will examine the idea of such an environment, and how it may or may not be reflected in our modern behavior. Now I want to consider a more basic question: What do we know about what human life was like during this period, and how do we know it?

Scientists have traditionally relied on three sources of information about early humans: (1) the fossils of those people and their associated artifacts, like tools and paintings; (2) the lives of modern-day hunter-gatherers—for example, the Kalahari bushmen, also called San—living what is thought to be a lifestyle closer to that of our ancestors; and (3) modern apes, particularly chimpanzees and their close relatives the bonobos, with which we share a common ancestor more recently than any other living animals. In recent years the scientist's kit has been expanded to include a new and potentially extremely powerful tool: the examination of our own genes, which bear the marks of past natural selection in ways we are only starting to appreciate. Each of these sources has advantages and drawbacks, and each feeds into our paleofantasies.

First let's look at the lives of what we now call modern hunter-gatherer or forager societies, or what used to be referred to as savages. Living in exotic and nearly inaccessible corners of the world, the Kalahari bushmen, the Hadza nomads of Tanzania, or the Aché of South America have sometimes been seen as a window into life before civilization. Early anthropologists classified human societies around the world according to their supposed evolutionary advancement, with hunter-gatherers in a state of arrested evolutionary development. Therefore, the reasoning went, studying those peoples would allow us to understand life before the advent of agriculture.

This classification, in addition to being objectionable from a sociopolitical viewpoint, is incorrect; all human groups have been evolving for the same period of time. But even after the idea that native South American Indians or other such human societies were in an earlier, and somehow more innocent or pure, state of nature was rejected, the notion that we can use them as models of what life was like for much of humanity's past has lingered.

If these people are hunter-gatherers and we know our ancestors were hunter-gatherers, why can't we look at these contemporary societies and draw inferences about our earlier way of life? The answer is that we can, but to a much more limited extent than many people would like. First, contemporary hunter-gatherers are variable in what they eat, how they divide labor between men and women, the way they raise their children, and a whole host of other features of daily life. Were our ancestors more like the Aché of tropical South America, who hunt small game but also eat a variety of plant foods; or the Inuit of the Arctic, who rely on large animals like seals for much of their food? They were probably like both, at different times and in different places, but it is impossible to tell at this stage which lifestyle was more common or which features were truly universal.

What's more, anthropologists increasingly doubt that any of the modern hunter-gatherer societies are truly "pristine," living the way that even their own ancestors did just a few centuries ago. The San, for instance, trade with their farming neighbors, and in fact may themselves have farmed sometime in the recent past. The Inuit are now known to have had complex networks of trade that, as anthropologist Rosemary Joyce of UC Berkeley says, "reached into Russia long before ethnographers came to describe their supposed simplicity."[20] In the 1970s, a tribe of hunter-gatherers called the Tasaday was supposedly discovered to be living as "remarkably peaceful remnants of Stone Age life," according to a *Science News* report by Bruce Bower.[21] But further investigation cast doubt on the authenticity of the Tasaday's isolation, and it now appears that at least part of the people's lifestyle, including their use of stone tools, was faked to attract media attention.

Many other contemporary foraging peoples are living in marginal environments far from their ancestral homes, pushed to the places where more technologically advanced people do not want to live. Or, like the Yanomami and other people living in the Amazon rain forest, they may be the remnants of much larger, and possibly quite different, populations that were decimated by Western diseases brought by the early European explorers. The Lacandon of Central America were once held up as a model of what "primitive" life would have been like if people had not begun to farm; we now know that they were pushed to the edges of the Spanish colony in the sixteenth century, adapting to their new environment by taking on behaviors like hunting and trading.

It isn't that one can't learn anything from these people—far from it, as they can often provide novel testing grounds for hypotheses developed in Western societies with different economies and cultures. Studying a diversity of cultures is the best way to understand when our generalizations are truly universal and when they inadvertently reflect biases from our own backgrounds. But the contemporary foragers do not mirror our past, and if they show admirable attributes, such as the absence of diseases like diabetes, or a lack of anxiety over unemployment, we cannot conclude that our ancestors were similarly blessed and that we have gone astray as we have taken on a settled existence in the course of evolution. As Joyce puts it, "We can still use the modern hunter gatherers to explore how small scale human societies work, but we have to see them as examples of small scale, not of earlier stages of evolution."[22]

The Neandertal makeover

Anthropologist John Hawks, who maintains a popular blog about human origins,[23] has a set of entries called the "Neandertal anti-defamation files."[24] According to the old conventional wisdom, Neandertals were those brutish hairy characters that lost out in evolution to the more cunning and—not coincidentally—more attractive *Homo sapiens*, who went on to make better tools, develop language, and generally evolve into, well, us. But ever since the

2008 finding of complex stone tools made by Neandertals, and more particularly since the 2010 discovery by Svante Pääbo and his colleagues that 1–4 percent of the genome of non-Africans today seems to have come from Neandertals, all parties have been hastily revising their stories.

Neandertals are suddenly cool;[25] they had bigger brains, and possibly more sex, than previously thought, at least according to the popular press. The *Guardian* newspaper sympathetically provided the headline "Neanderthals: Not Stupid, Just Different."[26] Wired.com huffed, "Neanderthals Not Dumb, but Made Dull Gadgets."[27] Scotsman.com was blunter: "Stone Me—He's Smart, He's Tough and He's Equal to any Homo sapiens."[28] At the new Neanderthal Museum in Krapina, Croatia, the displays even intimate that the Neandertals brushed their teeth. (Perhaps more disturbingly, or at least oddly, the former rocker Ozzy Osbourne attributes his survival of past excesses, such as drinking "up to four bottles of Cognac a day," according to the *Daily Mail*, to having Neandertal genes, a discovery he purportedly made with the help of a private genetics company.[29])

Since the Neandertals themselves, along with other now-extinct forms of humans in our direct lineage, are no longer around to tell us what they were like, we have to use what they left behind, in the form of tools, the remains of the animals they ate, and their own bodies, to understand the lives of our ancestors. How much can we conclude from these traces of bone and rock, and what do they tell us about where we came from?

Even just a fragment of skull can provide enormous amounts of information about the brain and body it was once part of. Any fan of the television show *CSI* knows that approximate age, height, and sex can be extrapolated from part of a skeleton, but modern anthropologists can easily outdo such fiction in their detective abilities. For example, the Neandertal brain, while about as large as that of a modern-day human, is shaped differently: it is more elongated and lacks the front bulge apparent in humans. Researchers at the Max Planck Institute for Evolutionary Anthropology in Germany closely examined the skulls of a Neandertal newborn baby and one older infant. The younger skull had a somewhat elongated braincase, like that of a human newborn, but the older Neandertal baby skull had not begun rounding in the areas at the top and base—the feature that gives human skulls their distinctive appearance. The researchers concluded that the Neandertals, even with their big brains, had a different trajectory of brain development than modern humans have, which could mean that our unique capabilities are a product of very early developmental processes.[30] In other words, it's not just the parts you have, but how they came to be, that determines your behavior. Other scientists are not convinced, given the scanty set of bones used for analysis, but regardless of the claim itself, the studies point to the amazingly detailed reconstructions that can be made using just a few bits of bone.

Teeth have long been used to draw inferences about diet, but recent developments in the analysis of dental enamel itself are providing an exciting new way to determine what our ancestors' diets were like. It's a rather more benign version of the idea that, when you have sex, you're also having sex with all of your partner's previous partners. When it comes to food, your teeth reflect not only the types of plants you eat, but the plants that the animals you eat ate themselves. Different kinds of plants use different kinds of carbon in their transformation of sunlight into energy, and those carbon variants can be tracked in the tooth enamel of animals, with trees and shrubs having a different chemical signature than grasses and sedges have. Anthropologists Matt Sponheimer and Julia Lee-Thorp examined the teeth of *Australopithecus africanus* and found evidence of appreciable amounts of the grass and sedge type of carbon, suggesting that these hominins ate seeds, roots, and tubers, and also could have eaten grazing animals.[31] They did not necessarily hunt big game; scavenging or eating insects like beetle larvae could have had the same result. Nonetheless, the study illustrates how much data can be gleaned from indirect sources.

What about reconstructing not brain growth or other aspects of anatomical evolution, but ancestral behavior? Here, too, fossils have played a role, but the extrapolation is a bit riskier.

"Neanderthals Really Were Sex-Obsessed Thugs," blared a 2010 headline from the UK newspaper the *Telegraph*,[32] and others were quick to follow, with AFP cheerfully noting, "Neanderthals Had a Naughty Sex Life, Unusual Study Suggests."[33] Neandertals were not the only subject of the study in question, but as I mentioned already, they seem to be undergoing such an image transformation that the usage was apparently irresistible. The supposedly salacious news comes from another examination of fossil fragments, finger bones this time, taken from Neandertals, four species of ancient hominoids, some early but anatomically modern humans, and data from four kinds of modern apes, including gibbons. The authors, led by Emma Nelson of the University of Liverpool, were interested in the kind of mating system our ancestors and their relatives might have had.[34] Did males compete with each other to gain access to multiple females, or were the species monogamous? Overall difference between male and female body size is often an indicator of the degree of multiple sex partners in a species, since selection for more competitive males often means that those males are heftier and hence better fighters. The more extreme the difference between the sexes, the more exaggerated the mating system. Male elephant seals, for example, are two to three times the size of females, and mating success in this species can be extraordinarily skewed, with a single bull in one population siring over 90 percent of the pups and the majority of the losers having no offspring at all.

Gauging overall body size difference between males and females can be tricky, however, if all you have are a few skeletal bones from each sex. What if you happened to measure an exceptionally large female or exceptionally small male? The finger bones were used because the ratio between the index (second) finger and the ring (fourth) finger is believed to reflect the levels of male sex hormones that an individual was exposed to while in the womb. In men, the second digit tends to be shorter than the fourth; in women, either the two fingers are more or less the same size or the second is slightly longer than the fourth. Lower ratios thus could indicate higher levels of male sex hormones, which might be expected in a species with greater male competition and less long-term pair-bonding.

Over the last decade or so, this ratio between the lengths of the digits has attracted attention from evolutionary psychologists, who purportedly found that sexual orientation, musical ability, female fertility, and several other characteristics were reflected in differences in the finger ratios of the subjects.[35] Nelson and her colleagues went a step further: instead of looking at the ratios within a population, they compared digit ratios among all of their specimens. We already know that gorillas, chimpanzees, and orangutans have multiple sex partners in a breeding season, whereas gibbons are relatively monogamous. Of the fossil species, all but *Australopithecus afarensis*—Lucy's species—had digit ratios suggesting that they were more like the gorillas or chimps and less like the gibbons, and the Neandertals were considerably more like those somewhat promiscuous species than are modern humans. Hence, presumably, the headlines about Neandertal "randiness," as the UK *Mirror* put it.[36]

Can we conclude, then, that the majority of our ancestors and their relatives, including the Neandertals, had a mating system in which men were able to have multiple wives, and that our monogamy is therefore a recent innovation? Maybe not.[37] Nelson and her fellow researchers were quite cautious, noting that their sample of bones was tiny, and that if the results were corroborated by, say, additional data on body size differences between the sexes, "digit ratios represent a supplementary approach for elucidating the social systems of fossil hominins."[38]

The use of digit ratios has been controversial almost from the start, with some researchers, including Lukáš Kratochvíl and Jaroslav Flegr of the Czech Republic, suggesting that the difference is a statistical artifact arising from the simple fact that men have larger fingers than women.[39] We also know virtually nothing about how to interpret the numbers themselves; the single anatomically modern fossil human, thought to be about 95,000 years old, had a ratio of 0.935, compared with the contemporary value of 0.957. The gibbons measured 1.009; the chimpanzees, 0.901. So if we are 0.052 units away from our monogamous relatives, but 0.056 units from our promiscuous ones, what does that make us? The answer is that we have no idea. John Hawks is skeptical about using the ratios to predict mating systems, partly because hands in apes and perhaps ancient humans are or were used in different ways, including locomotion, and hence are subject to different types of natural selection.[40]

I do not mean to argue that we should throw up our hands—whatever their digit ratios—in defeat and give up on the prospect of deducing anything about the ancestral human lifestyle from fossils. But the fossils themselves are often so limited, and the links between steps of logic based on so many assumptions (does prenatal hormone level always mean greater male competitiveness? do infant brain differences reliably translate into adult behavior?) that creating a clear picture of our past from our remains is not a straightforward proposition.

How chimpy are we?

No other species of the genus *Homo* are alive today, which means that if we want to use living animals to learn about what our ancestors must have been like, we have to look to somewhat more distant relatives. The resemblance between humans and the great apes, such as chimpanzees, has not escaped people's notice ever since those other primates were discovered, and it fostered the erroneous idea that we are literally descended from modern monkeys. The truth, of course, is that we and the monkeys and apes arose from the same ancestor, with humans splitting from the apes approximately 5–7 million years ago. That is still quite recent, which means that the traits we see in chimps, such as tool use, group hunting, or violent conflict between social units, can look like the remnants of our own heritage, and they have sometimes been seen as such.

Richard Wrangham and David Pilbeam titled a book chapter "African Apes as Time Machines,"[41] and Wrangham generally makes the case for humans exhibiting more "chimpiness" than similarity to any of the other great ape species.[42] Some of that chimpiness has to do with the apes' propensity for male dominance and violence—a propensity not shared by our other closest relative, the bonobo. Bonobos have more egalitarian societies in which social tension is resolved by sex, which occurs between members of the same sex as well as between males and females.

The study of bonobos began much later than that of chimpanzees, whose societies were documented starting with Jane Goodall in 1960, and some researchers—for example, Adrienne Zihlman[43]—have wondered whether our view of human nature might have been different had this order been reversed, with the pacifist bonobos used as models for our ancient selves. In such a scenario we could have formed a picture of early humans as naturally conciliatory, rather than pugnacious, and hence seen modern warfare as an aberration rather than a natural outcome of a violent heritage. Alternatively, when the discovery of the skeletons of *Ardipithecus*, that purported hominin from 4.4 million years ago, was published in 2009,[44] other anthropologists worried that the idea of studying an actual human relative, even a fossilized one, would sideline efforts to use modern species of apes to draw conclusions about early human lives, and hence needlessly limit the sources of information about our past.

So, can we use modern primates to understand our own evolution? The answer is yes, but not because the apes are time machines or living fossils or any of the other clichés often employed. Because chimpanzees, gorillas, and bonobos are so like us in appearance, and they have been since we all split from our common ancestor, it is reasonable to suppose that selection has acted on them as it has on other apelike primates, our closer relatives the hominins. For example, unlike many other mammals, we and the other great apes rely on vision and hearing, not smell, so our communication has evolved via sight and sound instead of odors, and the same was almost certainly true of our ancestors. But that is not the same as saying that the more recently we shared an ancestor with a species, the more similar we are to it in all respects. And it also doesn't mean that the apes themselves have remained identical to their 5-million-year-old ancestors; they have been evolving in their own environments just as we have.

One complication of assuming that the modern great ape is a frozen record of our shared past is that recent shared ancestry doesn't necessarily mean genetic similarity in any particular trait. It all depends on how quickly evolution occurred, and different traits evolve at different rates. In other words, suppose that our common ancestor was warlike, and for that reason chimps now are warlike. Even though modern human genes are very similar to those of modern chimps, if natural selection in humans has favored cooperation instead of conflict, we can retain that genetic similarity but acquire new traits of our own. How long a characteristic has been around doesn't argue for, or against, its existence in the first place. As Wrangham and Pilbeam put it, "But suppose, against all odds, that convincing evidence emerges to show that lethal raiding is 6 million years old in humans, whereas concealed ovulation [the absence of clear signs of periodic sexual receptivity, like heat in dogs and cats, a trait linked to monogamy and longer pair bonds] is 'only' 1.9 million years old. Should this matter to our sense of ourselves, that violence is 4.1 million years older than peace? Not at all."[45] Asking which traits changed quickly, and why, is a more compelling pursuit than trying to establish a baseline for our essential chimp—or gorilla, or bonobo—nature.

Were the universal people real people?

One of the many pleasures of travel to foreign lands is making connections with the locals, people who may be extremely different from us in their dress, eating habits, or religion but who often turn out to share some basic qualities—perhaps a love of children or a fear of snakes. Global conflicts over land or politics notwithstanding, people are people.

Anthropologist Donald Brown was sufficiently intrigued by this similarity around the world, noted not just by tourists but also by scholars, that he began to catalog what he called "human universals." Despite ethnographers' emphasis on cultural differences, Brown claimed in his 1991 book that "nowhere in the ethnographic literature is there any description of what real people really did that is not shot through with the signs of a universal human nature."[46] Brown was arguing against a purely cultural interpretation of what people do, suggesting instead that biology and evolution, interacting with the environment, have produced common behaviors in all human beings. Some of the universals include incest avoidance, the rough structure of language, a male-dominated political life, use of mind- or mood-altering substances, and the aforementioned fear of snakes. Brown's vision of these universals was remarkably detailed, not only about what people did but about how they felt: "Universal People . . . may not know how to make fire, but they know how to use it . . . Tools and fire do much to make them more comfortable and secure."[47]

If these universals are real, does that mean they reflect our ancestral behavior, so that we can use them as a way to extrapolate to our "most natural" modern behavior? Not really. Although I agree that the similarity among human groups around the world is striking, and I also concur that human behaviors are the result of evolution, it does not follow that we can use these similarities to construct, well, a paleofantasy of what our ancestors were like. Rosemary Joyce says that her archaeology students often want to support arguments about human nature and our own past with traits that are found in a diversity of modern societies. "But," she notes, "we could find that every human society living today had a particular social practice, and that won't tell us how human ancestors in the deep past acted."[48] The catch is that evolution is continuous, and we might all show similar patterns of behavior because we've all been subject to similar selection pressures.

How, then, did the cavemen live? It is not as clear-cut as the paleo proponents would like to believe. True, we know that early humans lived as hunter-gatherers, used stone tools to butcher their prey, produced art, and had a number of other attributes. And we can learn a great deal about how selection acts on social behavior by observing our close primate relatives. But there was no single Paleo Lifestyle, any more than there is a single Modern Lifestyle. Early humans trapped or fished, relied on large game or small, or collected a large proportion of their food, depending on where in the world they lived and the time period in which they were living. At any given moment, humans were doing some things that primates had been doing for 10 million years, such as using alliances among individuals to gain social status, and some things that were relatively recent evolutionary developments, such as making stone tools that could be attached to a handle instead of simply thrown or held in the hand. Neither one is more "authentic" than the other. Whether that is of interest to the Manhattanites trying to install meat lockers in their tiny apartments is another matter.

2

Are We Stuck?

Although we can argue about exactly how our ancestors from 50,000 years ago or more might have lived, it's undeniable that the time since people started living in settled groups larger than the size of a few extended families is extremely short. The ubiquitous clock metaphors for the history of the Earth have all of humanity shoved into the last few minutes, making agriculture and subsequent developments measurable only in nanoseconds. (In the Neanderthal Museum in Croatia, the time line of evolution is portrayed in a twenty-four-hour day, with "mankind's relations" appearing at just eight minutes before midnight.[1]) Even within the span of human evolution, the relative proportion of that period we have spent in agrarian settlements rather than as foragers or pastoralists is even more minuscule. And evolution is usually billed as a ponderous process, requiring thousands and thousands of generations before its effects are realized. It can seem logical, then, to assume that we have, as evolutionary psychologists and others are fond of stating, Stone Age genes ill suited to our Space Age lives and environment, and that we suffer the consequences. Or, as the Web page "Evolutionary Psychology: A Primer" puts it, "Our modern skulls house a stone age mind."[2]

Reasonable though this conclusion may seem, however, it is wrong, or at least it is correct only in such a broad sense as to be nearly useless. It is also often conflated with a different, though related, point, frequently made by paleo-diet proponents like Loren Cordain, that adopting agriculture sent humanity spinning down a starch-choked path of doom.[3] But in reality, we have two questions, both of which I will examine in this chapter. First, why, exactly, was this particular shift so momentous in the first place? In other words, what did agriculture do that was so revolutionary, and were all of the resulting changes bad? Second, given the consequences of humans giving up a less settled existence a mere handful of millennia ago, are we therefore stuck with the bodies—and minds—that we had before the transition to agriculture? Saying that agriculture and its concomitant changes to our diet and politics were bad for us, as Jared Diamond[4] and others do, isn't the same thing as saying we are trapped in an agriculture-induced cage—and an obese, sickly, socially stratified cage at that.

The curse or blessing—or both—of agriculture

Once the human species had spread out of Africa, people probably lived as hunter-gatherers in small groups until the rise of agriculture, which anthropologists Gregory Cochran and Henry Harpending call the Big Change.[5] No one denies that it was a major milestone, but several scientists go further and claim that it was the beginning of a downward spiral. In 1987, Jared Diamond, who later wrote such best-selling and influential books about the history of humans on Earth as *Guns, Germs, and Steel*, titled an article on the establishment of agriculture "The Worst Mistake in the History of the Human Race." In it he says, "With agriculture came the gross social and sexual inequality, the disease and despotism, that curse our existence."[6] An article in the British newspaper the *Telegraph* about Diamond's and others' work is similarly gloomily headlined "Is Farming the Root of All Evil?"[7]

Spencer Wells of the National Geographic Society goes even further: "Ultimately, nearly every single major disease affecting modern human populations—whether bacterial, viral, parasitic or noncommunicable—has its roots in the mismatch between our biology and the world we have created since the advent of agriculture."[8] And environmental writer and activist John Feeney pulls out all the stops with, "As hunter-gatherers, we blended gracefully into Earth's ecosystems. Then everything changed. Civilization is made possible by agriculture. Agriculture is unsustainable. If it weren't obvious already, you can see where this is going."[9]

The first point to clear up before we tackle all this pessimism is one of definition. Agriculture can be informally defined as growing one's crops and domesticating or at least keeping animals, rather than simply picking up what nature provides.[10] But anthropologists distinguish three kinds of such food production: horticulture, pastoralism, and intensive agriculture. People probably started out with horticulture, in which relatively unmodified crops are grown and cultivated with simple tools such as digging sticks. Modern-day horticultural societies include the Yanomami of South America, who combine growing manioc, taro, and some medicinal plants with foraging and hunting in the forest for the remainder of their food. Some horticulturalists today (and probably many in the past) spend part of their time as nomads, rather than living in permanent settlements. When they do form relatively sedentary groups, those groups are small, not likely to cluster in towns or cities.

Pastoralists, who rely on domesticated herds of animals that feed on natural pasture rather than on food provided by their keepers, have probably always been less common than crop-cultivating people, though even today a few groups, such as the Saami (known also as Lapps) of Scandinavia, who herd reindeer, persist. The animals are sometimes kept in one place for a few months at a time, as when the Saami keep female reindeer in corrals for milking during the summer. Although the reindeer, like other animals kept by pastoralists, provide the bulk of the Saami people's livelihood, the Saami and other pastoralists also trade with agricultural groups for other products, like plant foods.

Intensive agriculture is more like the form of growing food most often practiced today, though it is still seen in societies we would probably classify as "traditional," such as the rice-farming cultures of Southeast Asia. Fields are more permanent than those used by the horticulturalists, who may "slash and burn" the areas they cultivate, leaving them in between growing periods, sometimes for years, to regain nutrients in the soil. In contrast, intensive agricultural societies actively manage their fields with fertilizers and use them full-time. They also use more sophisticated tools, though these may simply be animal-drawn plows, not engine-powered cultivators. Crops are raised not only for eating by those who cultivate them, but for sale, which means that people can live in larger groups, with a division of labor between those who do the growing and those who buy or trade for the produce. That division of labor in turn means that resources—food itself or the means to purchase it—are not always divided equally, and society can become stratified.

Other than providing points of discussion to anthropologists, why do these distinctions matter? They matter because it is easier to accuse Monsanto-like agribusiness of causing widespread obesity and hypertension than it is to do the same thing to a few dozen people scrabbling in the ground for tubers using pointed sticks. And small-scale agriculture may have been around a great deal longer than people think; we are only now discovering that even the manly Neandertals had grain fragments between their teeth, and that early humans ground grains into flour, as I discuss in more detail in Chapter 5.

All of these gradations make it difficult to determine exactly when the woes associated with the shift to agriculture—increased levels of infectious diseases, reliance on one or a few food sources—first appeared. As archaeologist Tim Denham and his colleagues point out, "Early agriculture is not a demarcated 'all or nothing' lifestyle that can be clearly mapped across space and tracked through time."[11] As with all other processes in evolution, the move to a different way of obtaining food came about in fits and starts, with some human traits adapting well to the changes and others not so much. This irregular but realistic progression of events makes Cordain's contention that "the Paleo Diet is the one and only diet that ideally fits our genetic makeup. Just 500 generations ago—and for 2.5 million years before that—every human on Earth ate this way"[12] a little suspect.

This is not to deny the changes that took place as agriculture—intensive or otherwise—became established. Most obviously, the human diet changed to include and eventually depend on crops such as wheat, rice, and other grains, which meant that larger populations could be supported in one place. It also meant that the relative proportions of carbohydrates and proteins in the diet shifted toward the more reliable starches, though exactly how much is uncertain. Recent evidence from Neandertals and other fossils suggests, for example, that early humans may have eaten, and even processed, grain foods much earlier than had been supposed. Nevertheless, postagricultural diets not only relied more on carbohydrates, but were far less variable than the diets of hunter-gatherers. Estimates of the number of different kinds of plants eaten by many hunter-gatherer groups range from 50 to over 100, depending on the location of the population. Nowadays, in contrast, according to David Harris of the Institute of Archaeology at University College London, "a mere 30 crops account for 95% of plant-derived energy in the human food supply, over half of which is provided by maize, rice and wheat."[13]

Why might reducing the number of foods we eat be a bad thing? Eating a varied diet is not necessarily inherently virtuous, though certain micronutrients are probably best obtained from a variety of foods. But a varied set of crops does provide a cushion against some kinds of food shortages, in a not-putting-all-your-grains-in-one-basket way. The Irish potato famine, for example, came about because a fungal disease wiped out the potato crop that the peasants of Ireland relied on for most of their caloric needs. The disease, in turn, was able to have such devastating effects because almost all the potatoes had been selected to be genetically uniform, with the size, shape, and flavor that made them tasty and easy to grow. If one potato plant was susceptible, that meant they all were, and thus the entire crop could be decimated in one fell swoop. Reliance on just a few food plants makes us vulnerable to similar calamities, and it is an ongoing concern among scientists and farmers today. It is debatable, however, whether a return to a hunter-gatherer existence—even if feasible—is the best, or only, solution to this problem.

Working harder than a chimpanzee

One of the biggest bones of contention, so to speak, about hunter-gatherers versus agriculturalists is that the latter work too hard, in terms of both the time spent on subsistence and the intensity of the labor required, or at least they work harder than people who do not farm. Wells puts it this way: "As hunter-gatherers, we were a species that lived in much the same way as any other, relying on the whims of nature to provide us with our food and water."[14] And the whims of nature are presumably easier to cajole than the rocky soil or recalcitrant cattle of the farm. Agriculture, then, is sometimes seen as bad because it is just plain too difficult.

It is true that at least some hunter-gatherers spend less of their day "working," defined as engaging in activities necessary for subsistence, than do many farmers. Richard Lee's classic 1960s studies of the Kalahari desert people found that they needed two and a half days per week to collect enough food; adding activities such as toolmaking and other "housework" brought the total to an enviable forty-two hours per week.[15] Jared Diamond notes that the Hadza of Tanzania managed to keep their weekly work time down to fourteen hours or less.[16] Other estimates vary, and many of the calculations have been criticized by some anthropologists, who claim that the societies cited are not typical hunter-gatherers. But it seems reasonable to conclude that farmers, particularly those engaged in intensive agriculture, do indeed work harder than most foraging peoples.

The problem is that those foraging peoples are themselves still working pretty hard, at least compared to many other species. Anthropologist Hillard Kaplan and colleagues suggest that a hallmark of more modern humans was the ability to get hard-to-acquire foods.[17] They classify foods as *collected*, such as fruit; *extracted*, such as termites that are in protected underground nests or tubers that have to be dug from the ground; and *hunted*, which are foods such as deer or other prey that are caught or trapped.

Other primates, including chimpanzees, also eat foods that require some of the same kind of processing, and the chimps even hunt from time to time. But only humans focus on the extracted and hunted types rather than collecting what nature's whim provides. And we humans—even those in hunter-gatherer societies—need long years of training before we have the skills to net a fish or bring down an ungulate. Men of the Aché of South America, one of the best-studied contemporary foraging societies, do not peak in hunting ability, measured in the amount of meat collected per unit effort, until they are thirty-five years old. Collecting tubers is also no walk in the park; women of the Hiwi people of Venezuela become maximally efficient at foraging for roots between thirty-five and forty-five years of age. Acquiring these skills takes time, and lots of it.[18]

We can draw two conclusions from these statistics. The obvious one is that hunting and gathering is more than lolling around waiting for grapes to fall into your mouth or meeting up with your mates for an occasional fun-filled hunting trip. It may not be the workweek of a Wall Street shark or a nineteenth-century sweatshop laborer, but it is not the idyllic life we might have imagined. Less obvious, though, is that the amount of time one spends making a living is a continuum among animals, humans included. Why do we have a paleofantasy about the ancestral hunter-gatherer, when our even earlier relations, the apes, spend even less time foraging? Should we be yearning for the days before tool use? And how do we balance time against effort? Is it better to mindlessly munch grass, which requires little effort but takes a lot of time to down, one determined mouthful after another, or to spend less of the day fashioning a complex fish trap that may yield no catch? Choosing agriculture as the point at which we all started to go downhill because we began to work too hard is simply not defensible.

Farming sickness

Irrespective of what they are eating, intensive agriculture allows more people to be supported in a society. Having larger groups of people around, and having them be more or less sedentary, has several consequences. As Wells and many other authors have noted, one of the clearly undesirable effects of agriculture is the proliferation of new diseases, both infectious and noninfectious. Here, then, we can point to an unmitigated downside to settling down and farming: infectious diseases, those caused by pathogenic organisms such as viruses and bacteria, were able to spread because when people are in one place, their waste tends to stay put as well.

For example, cholera outbreaks occur when bacteria from infected feces contaminate the water supply, which is a problem only if you keep going back to the same polluted source to wash and drink. The disease can't establish itself in a continually moving population, so hunter-gatherers would not have suffered from it. Similarly, the virus that causes measles requires a fresh set of victims to be maintained in a population, so even if a small band of humans was infected with it, the disease would eventually have died out. In more densely populated areas, however, measles and diseases like it can be perpetually recycled into newly vulnerable targets.

Farming usually also means domesticating animals, and those animals can harbor diseases of their own, many of which are unwittingly passed to their caretakers. Worms, fungi, bacteria, viruses—our pets and work animals can be infected with all of them, and all have been implicated in human diseases as well. Smallpox, influenza, and diphtheria are all thought to have originated in nonhuman animals. Although wild animals can be a source of disease for hunters, the risk is much lower simply because the animals are not in contact with humans for very long.

Noninfectious diseases that would have appeared in newly agricultural human societies include vitamin deficiency diseases such as pellagra or scurvy, simply because agriculturalists tend to have fewer kinds of food in their diet, as I mentioned earlier, and the chance of relying on one or a few foods that lack essential nutrients is high. The skeletons of ancient farmers are filled with evidence of tooth decay, iron deficiency anemia, and other disorders. Diamond notes that the Greek and Turkish skeletons from preagricultural sites averaged 5 feet 9 inches in height for men and 5 feet 5 inches for women, but after farming became established, people were much shorter—just 5 feet 3 inches and 5 feet, respectively, by about 5,000 years ago, probably because they were suffering from malnutrition. The teeth from skeletons of Egyptians who died 12,000 years ago, about 1,000 years after their people had shifted from foraging to farming, were rife with signs of malnutrition in the enamel: a whopping 70 percent of them, up from 40 percent before agriculture became widespread.[19]

Then a funny thing happened on the way from the preagricultural Mediterranean to the giant farms of today: people, at least some of them, got healthier, presumably as we adapted to the new way of life and food became more evenly distributed. The collection of skeletons from Egypt also shows that by 4,000 years ago, height had returned to its preagricultural levels, and only 20 percent of the population had telltale signs of poor nutrition in their teeth. Those trying to make the point that agriculture is bad for our bodies generally use skeletal material from immediately after the shift to farming as evidence, but a more long-term view is starting to tell a different story. For example, Timothy Gage of the State University of New York at Albany examined long-term mortality records from around the world, along with the likeliest causes of death, and concluded that life span did not decrease, nor did many diseases increase, after agriculture. Some illnesses doubtless grew worse after humans settled down, but life has had its "nasty, brutish, and short" phases at many points throughout history.[20]

A deeper gene pool, with more unequal swimmers

Regardless of whether the people existing after agriculture were happier, healthier, or neither, it is undeniable that there were more of them. Agriculture both supports and requires more people to grow the crops that sustain them. Estimates vary, of course, but evidence points to an increase in the human population from 1–5 million people worldwide to a few hundred million once agriculture had become established. And a larger population doesn't just mean scaling everything up, like buying a bigger box of cereal for a larger family. It brings qualitative changes in the way people live.

For example, more people means more kinds of diseases, particularly when those people are sedentary. When those groups of people can also store food for long periods, the opportunity arises for sequestering that food, creating in turn a society with haves and have-nots. Many authors, including Diamond and Wells, have detailed the resulting social stratification, carrying with it increased division of labor and specialization, the growth of religion and government, and countless other marks of civilization, from architecture to money. Wells suggests that agriculture also fueled a change in human attitudes toward nature, from respect to a need for control, which in turn led to some of the planetary environmental problems of today.

Scientists and scholars ranging from the late Stephen Jay Gould to José Ortega y Gasset have lamented the supposed increase in warfare and violence as people moved from hunter-gatherer to settled life. The horrific large-scale violence we see today is sometimes attributed to the impersonal nature of warfare via airplanes and missiles, but the facts are hard to come by. Contemporary hunter-gatherers vary in the amount of warfare they exhibit, and as I pointed out in the previous chapter, because of modern influences on their behavior, any conclusions about our ancestors' violent predilections based on today's foraging peoples should be taken with a grain of salt.

Economist Samuel Bowles examined the percentage of adult mortality attributed to warfare from archaeological sites (where deaths from weapons can be determined using marks on the skeleton) and ethnographic records around the world dating from the nineteenth century, mainly before contact with modern industrialized society.[21] The data reveal a startling 14 percent of deaths from such violence, a much higher proportion than would be found in most societies today. Bowles goes on to suggest that selection at the level of the group could have increased the frequency of altruistic genes in early humans, because groups with sacrificing members would have been better able to persist in the face of frequent attacks. Whether or not Bowles is correct, his findings do not support a pacifist history of early humans that became bloody only after people became farmers. Psychologist Steven Pinker argues that human society has, in fact, become much more peaceful of late.[22]

More people, however, also means more genes. Not more genes in each individual, but more genes overall, as a simple result of there being more humans on Earth. While a larger population has obvious drawbacks, including overcrowding and high demand on resources such as clean water, it also has a sometimes overlooked benefit: more fodder for natural selection to act on. Evolution requires mutations, small alterations in the genes, to do its work. Beneficial traits, whether those are air-breathing lungs instead of gills or the ability to throw a spear, depend on new genes or combinations of genes, and the ultimate source of new genetic material is random mutation. Think of mutations as lottery tickets: the vast majority of them are losers, but the only way to increase your chances of winning is to buy a larger number of entries. As Cochran and Harpending point out, once the human population began expanding at a rapid clip, "favorable mutations that had previously occurred every 100,000 years or so were now showing up every 400 years."[23] And such favorable mutations spread more quickly in larger populations. Hence, a bigger population can evolve faster. John Hawks and his colleagues calculated that in the last 50,000 years, nearly 3,000 new adaptive mutations arose in Europeans.[24]

What this means is that the population explosion after agriculture, despite its well-known drawbacks, also carried some important positive changes that may have been overlooked. Cochran and Harpending also believe that human intelligence increased dramatically once groups became larger, via the same more-tickets-in-the-lottery mechanism.[25] Adam Powell and colleagues at University College London suggest that group size, not necessarily related to the birth of agriculture but among early humans in general, was key to the uptick in cultural and technological complexity seen during the Upper Paleolithic in many parts of the world.[26] Tools, weapons, art, and ritual objects all became more complex, and evidence of long-distance trading emerges in the archaeological record.

Exactly when that increase happened differs around the world, with a more rapid transition in Europe and western Asia than in northeastern Asia and Siberia. Why the variation? Powell and colleagues believe that it is much easier to lose skills like toolmaking if a group is small and only a few individuals possess the crucial knowledge. One unfortunate accident, and the village elder who transmitted information about the best place to get stones for spear tips, or a more sophisticated method for drawing on cave walls, is gone, and with that person the skills he harbored. Larger groups, or frequent migration between populations, provide some insurance against those skills being lost forever.

It is important to keep in mind that neither the benefits of human population growth, such as the flowering of genetic potential or cultural complexity, nor the more dismal consequences of agriculture, were directed. Spencer Wells looks at the advent of farming as akin to humanity diving off a cliff. Humans, he says, "divorced themselves—and us—from millions of years of evolutionary history, charting a new course into the future without a map to guide them through the pitfalls that would appear over the subsequent ten millennia."[27] He rues the "unintended consequences" of the establishment of agriculture.

The problem is, all of evolution's consequences are unintended, and there are never any maps. Arguably, apes, by moving from trees to plains, made their world spin just as out of control as we did when we began to grow crops. Either way, no one was aiming anywhere. As I discussed in the previous chapter, all those cute cartoons showing fish anxiously, or ambitiously, gazing up the shore toward ever-more-bipedal animals that eventually tote briefcases and wear Prada are just that—cartoons. Evolution is continuous, but it is not goal-oriented. It is not as if we were on a predestined path toward enlightenment when agriculture suddenly threw a plow into the works and made us deviate into obesity and disease.

Mired in the Stone Age, or in the EEA, or somewhere

Whether agriculture was a boon or a burden, what about the idea that because our bodies and minds evolved during the millennia before agriculture came about, we are still hobbled by our ancient genes in a world that has changed beyond recognition from our days as hunter-gatherers? The notion of a mismatch between modern and ancestral humans can be seen everywhere from diet books detailing how cavemen ate to conjectures about why powerful male celebrities seek out young women. As the quotations at the start of this chapter illustrate, we often look to our evolutionary past to explain the woes of our apparently ill-adapted present.

This idea of being stuck in a world to which we are not adapted is perhaps most elaborated on by the evolutionary psychologists. Evolutionary psychology, a field that purports to explain human behavior using evolutionary principles, relies on a concept called the Environment of Evolutionary Adaptedness, or EEA. The original idea of the EEA was developed in the late 1960s and early 1970s by psychologist John Bowlby, who was interested in how children become attached to their parents and vice versa. The EEA was later used as a linchpin for examinations of adaptation in the human mind. As Leda Cosmides and John Tooby, some of the leaders in evolutionary psychology, put it:

> Our species lived as hunter-gatherers 1000 times longer than as anything else. The world that seems so familiar to you and me, a world with roads, schools, grocery stores, factories, farms, and nation-states, has lasted for only an eyeblink of time when compared to our entire evolutionary history. The computer age is only a little older than the typical college student, and the industrial revolution is a mere 200 years old. Agriculture first appeared on earth only 10,000 years ago, and it wasn't until about 5,000 years ago that as many as half of the human population engaged in farming rather than hunting and gathering. Natural selection is a slow process, and there just haven't been enough generations for it to design circuits that are well-adapted to our post-industrial life.[28]

The EEA is the environment in which a particular characteristic, like the eye, or a love of sweets, evolved. It is not simply equivalent to the African savanna of 100,000 years ago, or any other single place or time. But because the human species spent so much longer on that savanna than it has in Midtown Manhattan, the evolutionary psychologists surmise that we simply have not had enough time to adapt to the modern environment. Tooby and Cosmides claim, "The key to understanding how the modern mind works is to realize that its circuits were not designed to solve the day-to-day problems of a modern American—they were designed to solve the day-to-day problems of our hunter-gatherer ancestors."[29] Hence the Stone Age mind (or genes). More simply, Edward Hagen says that the EEA "is the environment to which a species is adapted."[30] Thus, all organisms have an EEA, from fish to bacteria to elephants.

Before going any further, let me point out that I see nothing wrong with trying to explain the psychology or behavior of humans using an evolutionary framework. I do, however, find that people have a hard time viewing themselves dispassionately, and when it comes to explaining our own behavior, we have a regrettable tendency to see what we want to see and rationalize what we already want to do. That often means that if we can think of a way in which a behavior, whether it is eating junk food or having an affair, might have been beneficial in an ancestral environment, we feel vindicated, or at least justified. It's different from "my genes made me do it," but the end result—that we are trapped, perhaps regrettably, in a web of behavior that we inherited from our ancestors—is the same. What I am arguing in this book is that such an approach misses the real lessons of evolution, not only because it is specious to suggest that our genes dictate infidelity, but because that trap does not exist.

Furthermore, appealing to the EEA doesn't help. First of all, let's look at those creaky old genes. What does it mean to say that our genes are old, but our environment new? Our genes come from our ancestors, who got them from their ancestors, and so on ad infinitum, or at least "ad Precambrian-um." Some of our genes are identical to those of worms, chickens, and even bacteria, while others arose much more recently. A gene crucial to sperm production, called *BOULE*, is found in virtually all sexually reproducing animals and is 600 million years old, far preceding, of course, the time when humans were on the African savanna, or were even mammals.

Genes change when mutation provides the raw material and then natural selection or other forces, such as individuals moving to a new place, or sheer random chance, act on that material. But they change piecemeal, in fits and starts, and the rest of the genome is dragged along higgledy-piggledy. Organisms don't get to shed their whole set of genes in one fell swoop, like a pair of ill-fitting trousers, even during major transitions like the shift from water to land—or from ape to human.

New molecular techniques are allowing scientists to pinpoint which genes are evolutionarily conserved—that is, essentially unchanged as different groups split off from each other in history—and which are more recent. While it is true that more recently separated groups, like humans and apes, share more genes than do distant relatives, like humans and carnations, that relationship does not mean that those shared genes arose at any particular point, in our hunter-gatherer past or elsewhere, and now cannot catch up. As anthropologists Beverly Strassmann and Robin Dunbar point out, "From a genetic standpoint, the Stone Age may have no greater significance than any other period of our evolutionary past."[31]

Which genes change is also important. Much has been made of the proverbial 98 percent genetic similarity between humans and chimpanzees (the actual percentage changes slightly depending on which expert you consult or what metric is used, with biologist Roy Britten recently suggesting that 95 percent is a more accurate figure[32]). But the add-'em-all-up approach is not likely to yield any insight into what genetic differences, whether small or large, really mean. Anthropologist Jonathan Marks points out that we share perhaps a third of our genes with daffodils. It all depends what scale of measurement you use. "So from the standpoint of a daffodil, humans and chimpanzees aren't even 99.4% identical, they're 100% identical. The only difference between them is that the chimpanzee would probably be the one eating the daffodil."[33] Without going so far as to argue for the rights of flowering bulbs, as people have done for chimps and other great apes, Marks notes that it is difficult to know what to make of the similarity free of context. But Loren Cordain says, "DNA evidence shows that genetically, humans have hardly changed at all (to be specific the human genome has changed less than 0.02 percent) in 40,000 years."[34] This purported lack of genetic progress is used to support Cordain's prescription of a hunter-gatherer diet, before agriculture came along with its newfangled ideas about growing grain and living in houses.

Setting aside whether we are in fact 2 percent, 1 percent, or 5 percent different from chimpanzees, or whether our genes really are less than 1 percent different from those of our Pleistocene ancestors, what that 5 percent or 0.08 percent contains is crucial.

The vaunted statistics are often obtained by counting up the differences in the components of DNA between two populations, or two species. Because these components occur in a pattern of chemicals called bases, we often speak of DNA sequences. But simply comparing sequences tells little about the function of the DNA. Rebecca Cann, a human geneticist and anthropologist at the University of Hawaii, is skeptical about extrapolating from DNA to meaningful difference. She points out that while "it is true that it is difficult to find coding sequence differences between two modern humans, it is not true that therefore the ones that do exist are unimportant. And we won't be able to tell this just looking at the 'parts list.'"[35]

In other words, if all you had was an alphabet, you could easily end up concluding that *Hamlet* and the script for an episode of *The Sopranos* were the same thing, since they use exactly the same letters. Perhaps that idea is a little far-fetched, but I trust the analogy is clear. And when it comes to genes, the "parts list" is woefully inadequate. The big question is not how many genes differ between ape and human, or between today's human and our ancestors of 50,000 years ago, but *which* genes differ. Changes in the fine biochemical structure of DNA happen over time, simply by chance. Other changes occur because of selection on human characteristics such as language ability. But as eminent evolutionary biologist Sean B. Carroll says, "How can we identify the 'smoking guns' of human genetic evolution from neutral ticks of the molecular evolutionary clock?"[36] Using the alphabet analogy, he means that we need tools to help us distinguish Shakespeare from soap opera in a way that shows the difference in content, not just a difference in the number of times the letter "a" or "b" is used.

Carroll and other geneticists are now focusing their attention on regulatory and developmental genes, the ones that direct the rest of the show and determine when in the early growth of an organism its genes are switched on or deactivated. Much of the genome contains noncoding DNA, or genetic material that does not produce proteins. These sections can direct other genes, or simply clutter up the chromosome like jars of rusty bolts in a garage. Their functions, and the rate at which they seem to have changed in comparison with other genes, are a hot area of research in evolutionary biology. What they do not tell us, however, is that our genes are so similar to those of other organisms, or to those of our ancestors, as to render us stuck in the past, or that the number of changes per se is a valuable yardstick. Carroll puts it this way: "The rate of trait evolution tells us nothing about the number of genes involved."[37] But the converse is also true: knowing how many genes have changed doesn't tell us about how fast a trait has become altered.

What's more, whether old or new, human genes are also far from uniform, even after all this time. Although we are more similar to each other than are the members of a group of chimpanzees, human beings are still remarkably genetically diverse. Some genes, such as those involved in lactose tolerance, are far more likely to be found in people whose ancestry comes from some parts of the world than in people originating from other parts. Even within groups, the most casual scrutiny shows genetic variation in traits ranging from ear shape to the ability to taste bitter compounds. Such genetic variation among individuals is the fodder for evolution because it provides a menu of options for natural selection. If the environment changes, one or another of those menu items might be suited for the new conditions. This means that we still have an ample supply of genes that can evolve, and we are not simply dragging around a set of genes that were best suited for the Pleistocene.

What about the other argument supporting the need to use our EEA—that we spent far longer as hunter-gatherers in small bands than as cubicle workers in an urban sea? It does stand to reason that longer periods of time give evolution more scope to work, and by that standard, 10,000 years doesn't provide as much of an opportunity as 100,000, or a million. But sheer time simply isn't the only relevant variable. My students often complain that if they had just had more time for an exam, or to write a term paper, they would have done better. More time means more opportunity to work for them too, or so the student frantically clutching a test paper after the bell goes off would have me believe. The sad truth, however, is that some of those students wouldn't get the correct answer, or write an A essay, if you gave them from now until doomsday, or the next geological epoch. Time matters, and of course if I allowed only fifteen minutes for students to write a five-page paper, I would get shoddy work from everyone. But time isn't the only thing that matters.

The same goes for evolution and our ability to adapt to a new environment, whether that is agriculture or life on land instead of water. Large changes take a long time. Olives don't become petunias in a few generations. But how long does it take for them to become bigger olives? We no longer have to satisfy ourselves with generalities like "the time since agriculture is too short." We can look for the answers. The length of time required for a change in genes to become common in a population is a question we can now at least partly answer with data, as I'll detail in the chapters that follow. In the meantime, while it is true that, as Tooby and Cosmides point out in the title of a 1990 paper, "The Past Explains the Present," the present has not stayed still.[38]

Genes, peaks, and mismatch

As an alternative to the EEA, prominent anthropologist Bill Irons suggested a modification: the Adaptively Relevant Environment.[39] The Adaptively Relevant Environment is a set of environmental features, such as the amount of rainfall or the abundance of snakes, that is important to a trait, such as having a fear of reptiles. In an environment brimming with cobras, those who shun snakes are at an advantage. If the environment becomes reptile-free, and people instead run screaming from garden hoses, telephone cables, and other wiggly cylindrical objects, the trait is no longer adaptive.

Irons' notion does not rest on life in a foraging society, and hence avoids what he calls "Pleistocentrism in which all human psychological adaptations are tightly tied to the conditions of Pleistocene foraging societies."[40] He sketches several human behaviors, including incest avoidance and striving for high status, and then analyzes them using the concept of an Adaptively Relevant Environment, arguing that the former evolved as a mechanism that avoids the deleterious effects of inbreeding when close relatives have children. The evolution of incest avoidance thus required "a social environment in which close kin, siblings, parent, and children are in intimate contact during the critical period of the first two or three years of a child's life, and in which intimate contact is rare between nonkin or distant kin when one or both parties are in the critical age range of newborn to three years."[41] This somewhat pedantic mouthful boils down to having an aversion to sex with those one is raised with from birth or thereabouts—a situation likely to have been common both in foraging societies and more recent ones. We therefore do not need to invoke a particular way of life as a reason for the behavior.

Irons also notes the difficulty of defining the precise environment in which any adaptation, whether dietary, psychological, or otherwise, occurred, since humans, and the other hominins before us, did so many different things in so many different places during the hundreds of thousands of years before agriculture. He also points out that many environmental changes occurred more recently than the end of the Pleistocene and do not seem to have hinged on the transition to agriculture.

I do not find any particular fault with Irons' concept, but I am not sure we need a new framework for understanding the evolution of human behavior in addition to the usual principles of evolutionary biology. Traits in organisms, human or not, evolve in a particular environment, and although I agree with the evolutionary psychologists, and Irons, that understanding that environment helps us understand the adaptation, we may not need a brand-new dedicated term for it.

Perhaps it would be just as helpful to invoke a concept that has been in use by biologists in various forms since the 1930s, when the distinguished evolutionary biologist Sewall Wright imagined that populations and their genes could be viewed as if they were in a three-dimensional landscape, with hills and valleys.[42] The vertical axis, or height of the peaks, indicates the success or fitness of a group of genes. If a population on a mountaintop changes the composition of its genes, it is likely to move to a less successful point, and hence any small changes probably will be selected against. Conversely, a population in a valley is likely to improve with small changes. The entire fitness landscape may well contain peaks that are even higher than the one that a given population, even one on a mountain, is already on, but those peaks might have valleys between them. Hence, a population on a peak cannot move very easily to a higher one, whereas a population in a valley probably will get better no matter what direction it takes.

From the perspective of the EEA, the point is that we already have a way to think about inertia in evolution. Populations get "stuck," and it may be difficult for their gene frequencies to change without having their overall level of fitness—the degree to which they are suited to their environment—get worse before it gets better. But that is a different, and more nuanced, claim than the declaration that we arrived at the Pleistocene, or at a way of life with small hunter-gatherer bands, and will be unable to escape until millions of years pass.

This is not to argue that our modern lives are not sometimes, perhaps frequently, mismatched with our ancestral environment, or that we cannot use our past to inform our present. The evolutionary psychologists, among others, have reminded us that not all human behaviors are currently adaptive. It is extremely plausible, for example, that we crave sugar and not fiber because we evolved in an environment where ripe fruit was both nutritious and in short supply. Seeking it out meant gaining calories that in turn made it more likely the seeker would have enough nutrition to survive and reproduce, passing on his or her cravings. Nowadays, in a world full of processed sugar in everything from ketchup to Mars bars, this eagerness to consume sweets backfires, resulting in high rates of diabetes, obesity, and other woes.

Fiber is also good for us, yet we seem to lack that same enthusiasm for filling our diets with bran. Why wouldn't natural selection have instilled a drive to seek out high-fiber foods similar to the drive it instilled for sweet foods? The answer is simple: fiber was abundant in our ancestral environments, and no one had to do anything special to acquire it. People eating a diet similar to that eaten by hunter-gatherers can consume up to 100 grams of fiber per day, in contrast to the standard American intake of less than 20 grams, just because their food is all unprocessed. No one who craved the prehistoric equivalent of broccoli or bran muffins in the Pleistocene would have been at a particular advantage over those who did not.

Being mismatched, however, is different from being stuck. Instead of asking how we can overcome our Stone Age genes, let us ask which traits have changed quickly, which slowly, and how we can tell the difference.

3

Crickets, Sparrows, and Darwins—or, Evolution before Our Eyes

When we think about rapid change in a species, humans are often the first to come to mind, perhaps because we are so used to the idea that the modern world is very different from the one in which we evolved. But our anthropocentrism betrays us here, as it does in so many other places. Scientists are discovering more and more examples of evolution occurring in the span of just a handful of generations in animals large and small. What's more, some of those examples illustrate the practical implications of evolution, and why fishermen and farmers, not just scientists, should take heed of its findings.

My own firsthand experience with rapid evolution reminded me of taking our cat William with us in a U-Haul when my husband and I moved from New Mexico to Ohio. William was generally a stalwart and pragmatic animal, but like most cats, he greatly disliked riding in a vehicle, and he spent much of each day's trip howling in his crate, waiting for the horror to end. We would smuggle him into the motel room each evening, letting him out to use the litter box and eat. He would immediately dart under the bed, collect his nerves for an hour or so, and then emerge to go about his business.

This routine worked for the first two nights, but on the third we happened to stay in a room with a bed that had a frame extending all the way down to the floor. William exited the crate with his usual alacrity and headed for the bed. He stopped; there was no opening for him to creep into. He circled the bed. Still no hiding place. He circled it again. He leapt on top of the mattress, as if ascertaining that it was, indeed, a bed, and then tried again. No dice. Repeat. It was as if his brain kept cycling through the same set of incompatible conclusions: *It's a bed, so it must have an under-the-bed. But there is no under-the-bed, so it must not be a bed. But it seems to be a bed, so it must have . . . etc.* He couldn't resolve the cognitive dissonance except by tiring himself out by jumping on and off the bed and eventually taking refuge in sleep. We made it to our destination the next day, to a house with a proper bed, for which we were all grateful.

My own story of one sense contradicting another involves less jumping, but just as much befuddlement, and it also was my introduction to just how quickly evolution can happen in the wild. For many years I have been studying a species of cricket that lives in Australia and much of the Pacific, including islands such as Tahiti, Samoa, and the Marquesas. At least 150 years ago, the crickets were also introduced to the Hawaiian Islands, where they can be found in grassy areas like the lawns around buildings. They behave much as any other field cricket does: males call to attract females, and females select mates on the basis of which songs they prefer. Unfortunately for the crickets, in Hawaii—though not their other haunts—they must contend with parasitic flies that can also hear the chirping. A female fly is incredibly good at locating a male cricket using his song, and when she does, she dive-bombs him, leaving in her wake not missiles but something far worse: her tiny voracious larvae. The maggots burrow into the cricket and develop inside his still-living body, gradually consuming his flesh. After about a week, the maggots emerge from the withered, dying shell of their cricket host, to pupate in the soil and finally buzz away as adult flies.

A field cricket with Ormia ochracea, *the parasitic fly that locates male crickets by their songs. In Hawaii, these flies seem to have spurred extremely rapid evolution in crickets, rendering them silent.* (Courtesy of Norman Lee)

This gruesome life history is remarkable for many reasons, but as a specialist in the evolution of the signals used in animal sex—peacock tails, cricket chirps, or elk bellows—I became interested in the dilemma it poses for the male cricket. Obviously, the more you sing, the more likely you are to attract a mate—an evolutionary jackpot for any animal. Singing is more or less a male cricket's raison d'être. But in this case, the more you sing, the more you expose yourself to the deadly parasites, and dead crickets cannot sing at all (a truism I have always thought would make a good title for a mystery novel). Selection thus operates in precise opposition, acting both for and against the same trait, and I have spent much of the last two decades trying to figure out how the crickets resolved their evolutionary conundrum.

My colleagues and I discovered many interesting parts to the solution, but the real stunner happened during what I thought would be a somewhat rushed and routine visit to Kauai. We had been doing our research on the Big Island, Oahu, and Kauai, and for several years I had been catching and dissecting crickets in each place to see how the males infested with parasites might differ from the ones that had escaped. It was always a small thrill to open up the body cavity under the microscope and see, in addition to all the normal cricket organs and tissue, a plump white maggot glistening under the cold fiber-optic light we use for dissection. The tiniest maggots have little black racing stripes along the sides, giving them a cheerful perkiness that belies their more sinister character.

None of my colleagues saw the charm of any of the fly larvae, large or small, but we all could see that the percentage of males harboring the maggots was always highest on Kauai. Nearly a third of all the males we caught were infested, and we wondered for years whether the relentless hunting by so many flies would eventually wipe out the cricket population entirely. We don't know how long the flies have been on the islands, but we know the crickets did not evolve there, so the relationship is relatively recent, suggesting that it might not be stable.

Indeed, starting in the late 1990s, it became increasingly difficult to find any crickets at all on Kauai. A field that once held dozens grew more and more silent, and by 2001 we heard only a single cricket calling in our usual field site. That doesn't mean others weren't around, of course, but the relative silence was symptomatic of a major decline in the population.

Thus, when my husband and I returned to Kauai in 2003, I was not optimistic. It was certainly possible that the flies had finally proven the death knell for the crickets on the island. And sure enough, when we drove up to the field where we usually found our samples, the night was silent. But having come so far, I figured we might as well get out of the car and take a look. We put on our headlamps and started walking up the road.

And that's when I felt like William the cat. Because in front of me, on each side, hopping on the ground and perching on the grass, were crickets. Lots of crickets. More crickets than I had seen on Kauai for years—maybe ever. I caught one. It was one of our species all right, and it was even a male, easily distinguishable from the females by the absence of a long, straw-like ovipositor used to lay eggs. But I didn't hear a thing.

Remember that singing is, well, what crickets do. Except for a few oddball species that definitely did not include my study subjects, the definition of a male cricket is an animal that gets out there every night, lifts his wings, and rubs them together to make noise. Individuals may refrain from calling sometimes for one reason or another—it's too cold, they are injured, or they mated so recently that they cannot produce sperm for the next female—but chirping is part of the cricket identity. So my mind started going through puzzled loops, much as I imagine William's had done years before: *These can't be crickets, because they aren't calling. But look, there are lots of crickets. Maybe they are a different kind of cricket that just suddenly appeared here on Kauai? Nope, they are my kind of crickets. But these can't be crickets, because they aren't calling. Maybe I've gone deaf? No, that's not it. But these can't be . . . etc.*

I would like to be able to say that I resolved the disharmony of the situation more quickly than William had done in giving up his futile perusal of the motel bed and falling asleep, but the truth is that although I shook myself out of my own cognitive dissonance and actually tried to use science to solve the puzzle, it took a few months before we had the answer.

The crickets, it turned out, weren't silent because they could have called but chose not to; they were silent because they lacked the apparatus to produce any sound at all. In the space of fewer than five years, or about twenty generations (taking a conservative estimate that the mutation appeared a few years before we noticed it, and with the crickets producing three to four generations per year), a new form of the cricket that we dubbed "flatwing" had become so common that we now estimate only about 10 percent of the crickets can still sing. (We just happened not to hear any on that first night, but later visits turned up a valiant handful of callers.) Additional research in my lab showed that the flatwings have a mutation in just one gene, but that single gene changes their wings so that they lack the equivalent of a fiddle and bow for making music.

Ordinarily, of course, not being able to call would be a tremendous liability for a male cricket, and if the mutation arose under normal circumstances, it's virtually certain it would be an evolutionary dead end for its bearer, who would be unable to mate and hence unable to pass on any genes. But the flies change the rules of the game, and the flatwings are protected from detection by the lethal parasites by their silence, a veritable cloak of inaudibility that provides an enormous advantage.

This leaves one major question: How can a female cricket find her silent mate? The answer is turning out to be complicated, with the females apparently willing to mate with the silent males as long as the males are near one of the few remaining callers. But for our purposes, the point is that the crickets are an example of one of the fastest cases of evolution in the wild, taking not hundreds or thousands of generations, but a mere handful. In human terms, twenty generations is only a few centuries.

Although my crickets evolved more quickly than many species, they are by no means alone in changing during a relatively short period. Contrary to the commonly held notion that evolution is a ponderous process, requiring geological spans to produce any detectable change, scientists are now discovering that, as Andrew Hendry and Michael Kinnison note, "The fundamental conclusion that must be drawn is that evolution as hitherto considered 'rapid' may often be the norm and not the exception."[1]

The birds who came in from the cold

Perhaps because people are always interested in how fast events can happen, and also because people seem to have differing views on how long Earth and its inhabitants have been around, the rate of evolution has piqued our interest virtually since Darwin. One of the first to suggest a speed record for genetic change in a population, the most basic definition of evolution, was an extravagantly mustached scientist named Hermon Bumpus. Like the other residents of Providence, Rhode Island, at the end of the nineteenth century, Bumpus, an assistant professor of comparative zoology at Brown University, experienced some record-setting bad winter weather. But unlike most of them, he decided to profit from the misfortune of others in the name of science.

Somehow—his 1899 paper is mute about the source—the day after a particularly severe storm, Bumpus was brought 136 dead or stunned house sparrows.[2] I have searched in vain for the person or persons who thoughtfully provided Bumpus with his subjects, and their reason for doing so; was he well known for a general interest in dying birds? Did his friends and neighbors know that he had wanted all his life to record the miseries of house sparrows, those English invaders of North America? Was the purveyor of the bird bodies also interested in natural selection, only to be summarily dismissed from Bumpus's subsequent work and denied a coauthorship? You don't just cart around 136 sparrows in your pockets in the hope that they will come in handy sometime.

Alas, the backstory to this experiment seems destined to be shrouded in mystery, but regardless, once Bumpus saw that about half the birds recovered after they were warmed, while the others died, he knew he had a gold mine on his hands. Natural selection requires some individuals to survive and reproduce while others fail, and here was an opportunity to see whether the sparrows that gasped to life in his laboratory had any characteristics that the others did not. So Bumpus methodically measured the size of bills, wings, legs, and other body parts in the survivors and those not so fortunate, and then compared the two groups. They differed substantially; birds that were either unusually large or unusually small fared worse than those clustered around the average size of the group, and Bumpus postulated that stabilizing selection, a kind of natural selection that winnows out the extremes and favors those in the middle, had been at work.

His results have been scrutinized many times since he published them, and scientists still argue about the best way to analyze his measurements, but the fact remains that Bumpus's sparrows are among the earliest examples of evolution occurring before our eyes, at least within the local population of house sparrows. Bumpus had not only confirmed that evolution happens in the wild; he had shown that it could, literally, occur overnight.

How many darwins does it take to screw in a lightbulb?

After Bumpus, several other scientists tried their hand at calculating rates of evolution, and at coming up with a unit of measurement that could be applied to it. It's fine to say that the sparrows' size changed after a storm, or that the silent crickets came to predominate after just five years, but how do you compare, say, a 10 percent increase in leg length after thirty generations with a 40 percent decrease in swimming speed after a hundred generations? Which is faster?

The scientist who first tried to formalize measuring the rate of evolution was the great British geneticist J. B. S. Haldane. Born into an aristocratic family, Haldane became an enthusiastic scientist at an early age, performing physiology experiments on himself (he once drank hydrochloric acid to see how it would affect his muscles), as well as a public intellectual, writing a dystopian science fiction novel published in 1924, and an avid Marxist. Haldane's paper outlining the technique for assessing how fast evolution occurs is regrettably much less colorful than his fiction, but it does suggest using the percentage of change in a characteristic—for example, leg length—and combining that measure with the characteristic's standard deviation, a statistical measurement of how much a trait varies within a population.[3] He suggested a unit of measurement called, appropriately enough, a "darwin," which means that if we had lightbulbs being installed at two different times, we could, at least in theory, ask the question in the title of this section.

Haldane's method, however, is best applied to changes that take a long time, like the interval between the appearance of different kinds of teeth in fossil horses from different time periods (the example he used in his 1949 paper). Long after Haldane's death from cancer in 1964 (while ill he wrote a poem called "Cancer's a Funny Thing"), paleontologist Phil Gingerich proposed a different unit, which he dubbed the haldane.[4] Both measures are currently used by scientists, although neither has made it into the popular vocabulary.

If that is how evolution is measured, what do we mean when we say that evolution is fast? These days, evolution is considered rapid if a population shows a genetic change over tens of generations or fewer, or sometimes as much as a hundred generations. Rapid evolution is sometimes called "contemporary evolution," to emphasize that it happens within a modern time span, or "evolution in ecological timescales," to emphasize that evolution can be important to occurrences within the lifetime of animals or plants, while they are undergoing ecological events like being eaten by a predator or attacked by a parasitic fly.

Beaks of eagles, and finches

In his 1936 poem "The Beaks of Eagles," Robinson Jeffers rather sanctimoniously chided that man needs "to know that his needs and nature are no more changed in fact / in ten thousand years than the beaks of eagles."[5] It is probably just as well that Jeffers found eagles a more suitable subject for poetry than finches, given the latter's place in the rapid-evolution hall of fame. Of course, we don't know for certain that eagles have been all that unchanging, which casts doubt on the whole "needs and nature" of man idea as well. But the finches, at least the ones that live in the Galápagos Islands off the coast of South America, have certainly been changing, in beaks as well as other body parts, for at least the last several decades.

The finches of the Galápagos, also known as Darwin's finches because they contributed substantially to his theory of how several different species can evolve from a common ancestor, have been studied by biologists Peter and Rosemary Grant since the 1970s. When I was in graduate school at the University of Michigan, where Peter held a professorship until he moved to Princeton, the "Finch Group" was a well-oiled machine, with students trooping down to the islands to measure and monitor the birds almost year-round. I was friends with several of the students whom Peter advised, so I got to exchange letters with them and listen in on the discussions of the field trips after the students returned, improbably tanned, in the Ann Arbor winter.

The first thing these interactions did was disabuse me of any notion that field work is glamorous (a valuable revelation, given my later predilection for crawling around on my hands and knees in the dark catching crickets). I got lots of letters about the monotonous diet, with moldy and weevil-ridden oatmeal receiving a particular mention, and the routine of getting up each morning on a tiny nubbin of rock, finding the birds, noting their stage in the nesting cycle, and keeping track of which birds had how many chicks and when. I also heard about measuring. A lot of measuring. The Grant crew measured the height and number of shrubs, the size of the seeds the shrubs produced, the width and length of the bills, legs, and wings on the birds that ate the seeds, and pretty much anything else they could lay their hands, or at least their calipers, on.

While the students were making their treks to the Galápagos in the 1980s, the area experienced a phenomenon called El Niño, an oceanographic event that drastically alters the usual rainfall patterns. Ordinarily the Galápagos Islands are dry, even desolate, with brief rains that fall during only part of the year. But between November 1982 and July 1983, an El Niño brought record-breaking rain, and the plants responded with flourishing growth followed by a bonanza of seeds. The Grants and their students were gobsmacked by the effect of El Niño, showing an enthusiasm that I must admit I did not entirely share; the lab meeting after they brought back the first photographs showing evidence of the rains involved scrutinizing each bush and rock on each slide, with detailed commentary along the lines of, "Did you see that big branch on the top of that shrub on the left? It's incredibly green, isn't it? I remember when that whole branch was brown, don't you? But it's green now!" After listening for half an hour or so, I got up and crept out of the room, unnoticed.

The rains and their subsequent effect on the foliage did, however, help to demonstrate evolution in action. The lists of measurements made their way back to the lab in Ann Arbor along with the students, and when the data on the birds were matched with the information on long-term weather patterns, an astonishing result emerged. Before El Niño, when the climate was relatively dry and fewer shrubs produced seeds, finches that were larger, with bigger bills, had survived and reproduced better than their daintier companions, probably because a large bill enables a bird to crack large, hard seeds more easily and the smaller, easier-to-eat seeds had all been consumed. Afterward, smaller individuals had their chance, until the more usual dry seasons came around once again. Over the three decades that the Grants and their coworkers studied the finches, the average body and beak size of one of the finch species, the medium ground finch, first became smaller, then rapidly became larger, and then decreased again, but more slowly, as the conditions favoring large or small bills and bodies changed and selected for different traits.[6]

What this means is that populations, and species, can and do change rapidly, over and over again, as the forces of nature change around them. They don't always change in the same direction, because evolution is more of a drunkard's walk than a purposeful path, as I discussed in Chapter 1. This means that the future success of any single individual depends on the circumstances into which it is born; a female ground finch that hatched in 1983, after the El Niño had passed, had to live for six years, and produce ten chicks, if she was to be guaranteed to replace herself and her mate with two of them. Most of her effort would have been in vain, with her offspring dying when they could not find enough food. But if she had been hatched just before the rains, in 1978, she could have been assured of leaving those two replacement young in just two and a half years, with only five babies. Either way, it makes you wonder whether Jeffers would have found consolation in the finches.

One fish, two fish, old fish, new fish

To continue in a literary vein, albeit a somewhat more lowbrow one, the poet Ogden Nash observed in "The Guppy" that while we have special names for many animal youngsters—cygnet, calf, cub, kitten—"guppies just have little guppies."[7] Indeed they do, and quickly; maybe we don't bother with names for baby guppies because they become adults in a matter of a few months. A female guppy can be sexually mature at two months of age and have her first babies just a month later. This unstinting rate of reproduction makes guppies ideally suited for studying the rate of evolution, and David Reznick, a biologist in my former department at UC Riverside, has been doing exactly that for the last few decades.

People usually think of guppies as colorful aquarium fish, but they also have a life in the real world, inhabiting streams and rivers in tropical places like Trinidad, where Reznick has done his fieldwork. As with the finches, guppies can experience different kinds of conditions depending on the luck of the draw, though for the fish it is the presence of predators, not the amount of rainfall, that makes the difference between who lives and who dies. A lucky guppy is born above a waterfall or a set of rapids, which keep out the predatory fish called pike cichlids found in calmer downstream waters. As you might expect, the guppy mortality rate—that is, the proportion of individuals that die—is much higher in the sites with the rapacious cichlids than in those without them.

Reznick has shown that if you bring the fish into the lab and let them breed there, the guppies from the sites with many predators become sexually mature when they are younger and smaller than do the guppies from the predator-free sites. In addition, the litters of baby guppies produced by mothers from the high-risk streams are larger, but each individual baby is smaller than those produced by their counterparts.[8] The disparity makes sense because if you are at risk of being eaten, being able to have babies sooner, and spreading your energy reserves over a lot of them, makes it more likely that you will manage to pass on some of your genes before you meet your fate. Reznick and other scientists also demonstrated that these traits are controlled by the guppies' genes, not by the environment in which they grow up.

How quickly, though, could these differences in how the two kinds of guppies lived their lives have evolved? Because there are numerous tributaries of the streams in Trinidad, with guppies living in some but not all of them, Reznick realized that he could, as he put it in a 2008 paper, "treat streams like giant test tubes by introducing guppies or predators" to places they had not originally occurred, and then watch as natural selection acted on the guppies.[9] This kind of real-world manipulation of nature is called "experimental evolution," and it is growing increasingly popular among scientists working with organisms that reproduce quickly enough for humans to be able to see the outcome within our lifetimes.

Along with his students and colleagues, Reznick removed groups of guppies from their predator-ridden lives below the waterfall and released them into previously guppy-free streams above the falls. Although small predatory killifish occurred in these new sites, these do not pose anything close to the danger of the cichlids. Then the scientists waited for nature to do its work, and they brought the descendants of the transplanted fish back to the lab to examine their reproduction. After just eleven years, the guppies released in the new streams had evolved to mature later, and have fewer, bigger offspring in each litter, just like the guppies that naturally occurred in the cichlid-free streams.[10] Other studies of guppies in Trinidad have shown evolutionary change in as few as two and a half years, or a little over four generations, with more time required for genetic shifts in traits such as the ability to form schools and less time for changes in the colorful spots and stripes on a male's body.[11]

More members of the biological rapid-response team

Although guppies have been particularly well studied, more and more animals are turning out to be capable of extremely quick evolution. The mounting examples are significant not simply because they are fascinating in their own right, but because, taken together, they put to rest the notion that evolution requires millennia. It turns out that even relatively complex traits can evolve quickly—a discovery that has led to increased study, and new understanding, of the way genes themselves interact. This research can in turn be applied to humans, enabling us to understand which of our own traits are likely to evolve quickly.

Most populations of blackcap warblers, European songbirds with males bearing the rakish head ornament of their name, sensibly spend the winter in southern Europe or northern Africa. But starting in the 1960s, a handful, and then a steady stream, of blackcaps could be found migrating to England, where they survive in people's gardens. Peter Berthold and his colleagues took some of the Britain-wintering blackcaps into aviaries and bred them; their offspring stuck to the new migration route, and further work showed that the birds' migration routes are inherited.[12] It would seem that the migratory behavior itself evolved in just thirty years.

Other examples of rapid evolution include quick changes in cold tolerance by sticklebacks,[13] small fish that occur in both marine and freshwater bodies of water, and a shift to longer straw-like mouthparts in the endearingly named soapberry bug when it encountered an invasive weed with fruits that held deeply buried nourishing sap.[14] In about a dozen generations, anoles, those small lizards sometimes erroneously labeled chameleons in pet shops, changed their body shape and the length of their hind legs when researchers put them in new places within their native Bahamas, with different competitors for food and places to hide.[15] The spires of snail shells evolved new curves in less than twenty years.[16]

A recently discovered, and rather clever, case of rapid evolution concerns some of everyone's favorite animals: toad-eating snakes. Okay, that is not exactly correct. More accurately, they are snakes that sensibly *avoid* eating toads, since toads, like other amphibians, are quite toxic. Poison dart frogs get their name because indigenous South Americans used the secretions from their skin to make their arrows or spear tips more deadly. Even if less lethal, other types of frogs and toads should generally be handled with caution, and you are well advised to wash your hands thoroughly after touching them.

Snakes, in contrast to frogs and toads, have skins that are dry and supple, without the nasty secretions of amphibians, but many species are susceptible to the toxins if they eat the frogs or toads themselves. Many kinds of snakes are not particularly discriminating about their prey and are what is called "gape-limited," meaning that they will strike at, and try to swallow, anything living that will fit into their mouths. Unfortunately for the snakes of Australia, for the last several decades, some of the likely candidates have been cane toads.

Cane toads were introduced in 1935 to Queensland to control pests in sugar cane fields. Like several other introduced predators, they turned out not to be particularly adept at their intended task, but they were extremely skilled at something else: making more cane toads. A single female toad can lay between 20,000 and 50,000 eggs when she spawns, far more than any native Australian amphibian, and she can breed more than once per year. In addition to poisoning their predators, the cane toads likely compete with native frogs and are thought to be able to upset the ecological balance of the wetlands where they occur, simply by eating far more snails, insects, and other invertebrates than the native animals would.

The government and various environmental organizations in Australia, as well as a lot of academics, have become quite concerned about the effect of cane toads, particularly because they have expanded mightily since their introduction and show no signs of slowing down: depending on the part of Australia you look at, cane toads are spreading at a rate of 5–50 kilometers (about 3–30 miles) per year. Efforts at toad control have not been met with resounding success, and they are difficult to implement, particularly when the public is invited to participate; a 2011 Queensland

"10,000 Toads Project,"[17] with its goal of culling that many of the animals in a single effort, was met with disapproval by the Royal Society for the Protection of Animals because some citizens proposed using golf clubs to kill the toads.[18] (RSPCA guidelines suggest placing the offending animals in the freezer as a more humane way to dispatch them.) A website associated with the threat of cane toads to northwestern Australia mentions a Cane Toad Working Group,[19] which, despite what you might think from the name, is not a chain gang of amphibians but a collection of concerned citizens and government officials who compare notes and collaborate on strategies for toad eradication.

But back to the snakes, and to rapid evolution. In addition to watching in alarm along with their fellow Aussies as the cane toads spread, Ben Phillips and Rick Shine at James Cook University in Queensland and the University of Sydney, respectively, started thinking about the toads' effect on the local snakes. If snakes can attack and swallow only prey that are smaller than the size of their open jaws, then species with small heads relative to their bodies will be less able to engulf a poisonous cane toad and thus more likely to survive and pass on their genes. The scientists reasoned that the toads would therefore have caused the snakes that were vulnerable to the toads to evolve different head sizes, making the snakes of today different from those that occurred before the cane toads were introduced.

Phillips and Shine then turned to the snakes preserved in jars at the University of Queensland and measured the heads, jaws, and bodies of four species of snakes collected over eighty years, spanning the time before and after cane toad introduction. Two of the species were predicted to be unlikely to have suffered any ill effects from the toads, either because they are too small to be able to eat a toad and ingest a fatal dose of the toxin or because they already possess, as some snakes fortuitously do, a physiological resistance to the toxin's effects. The other two snake species were more likely to encounter and try to eat the toads. Phillips and Shine predicted that the toad-vulnerable species now would look different from what it looked like before the cane toads arrived, whereas the other species would not.[20]

They were right. The differences were not enormous, but they were statistically detectable, and they followed the level of exposure to the cane toads in the places where the snakes had been collected. Over just a few decades, the affected snakes had evolved smaller heads and differently shaped bodies.

The cane toads themselves are not immune to rapid evolutionary change. Phillips, Shine, and their colleague Greg Brown also measured the toads at the edge of the invading population, that front of colonizing individuals responsible for the steady spread of their kind. The toads are not, of course, consciously trying to be imperialistic and claim new ground for all cane toad–dom. Instead, some individuals are more likely to move away from the place they were hatched, and because the cane toads have many characteristics that make them successful invaders, such as the abilities to eat many kinds of foods and to tolerate a wide range of temperatures, those itchy-footed individuals will be the ones that make it at the forefront. Not only is the predilection for moving far from your parents' home genetically inherited, but, according to Shine, "invasion front toads are longer-legged, more active and more mobile than their cousins back in long-established populations."[21] Of course, since once you've moved away from home you are unlikely to date anyone from the old neighborhood, the fastest-hopping toads end up mating with other sprightly individuals, and the process accelerates.

This Hun-like power does come at a price. According to Terri Shine (designer and writer for the web site CaneToadsinOz.com), "That incredible increase in toad speed has put enormous pressure on the toads' bodies—bodies that had evolved to sit around a swamp in Brazil and eat flies, not sprint across Australia like an

Olympic long-distance athlete."[22] The recently evolved toads are prone to severe spinal arthritis, making hopping more difficult and probably painful, though they seem to persevere regardless. In time, it is possible that toads better able to cope with all that frenetic (for a toad anyway) movement will replace the old creaky ones, but for the moment, as Shine says, "Cane toads really are tough!"[23] And it all happened in less time than it often takes for a new political party to come to power.

Who has what it takes?

Why do we see rapid evolution in some species and not others? Genetic change within a handful of generations may be more common than we had previously thought, but it is still not the rule. So, can we predict which organisms are likely to evolve quickly, and under what circumstances? This information turns out to have practical applications, because it means we can understand how both farmed species and wild ones used by humans—for example, by hunters—will respond to our interference in their lives.

Although the species that have demonstrated rapid evolution vary enormously, from water fleas to birds to, of course, toads, a hefty proportion of them have been fish. And not just any fish: the guppies, certainly, but also several kinds of salmon, cod, whitefish, and herring. Aside from the guppies, these species share a common characteristic: they are fished by humans. Bighorn sheep are also showing marked recent changes in their appearance, and they are hunted by people as well.

We have known for a long time that hunting and fishing can reduce the numbers of wild animals, or even cause their extinction. Passenger pigeons virtually covered the sky over much of North America until their demise in the wild in 1900, and by 1992 the cod fishery off the coast of Canada and New England was about 1 percent of its size in the preceding centuries. Even a less dramatic form of hunting, the use of insecticides, has made pests like boll weevils or malaria-carrying mosquitoes less common, at least in some parts of the world. But what about the individuals that are left behind?

It turns out that the surviving individuals often are different from those that were removed from the population. Hunting and fishing can act at least as powerfully as any winter storm or seasonal drought in winnowing out certain body shapes, and certain genes, from a population. If trophy hunters prefer to shoot the rams with the largest horns, those males may not end up breeding as much as they would have in an unmolested population, which means that the smaller-horned males are more likely to pass on their genes. Indeed, horn size seems to have decreased in some bighorn sheep populations, along with antler size in some Canadian moose.[24] Other shifts have not been so obvious; just as the predators in the streams below the waterfalls caused the guppies to mature earlier and have more and smaller offspring, widespread fishing seems to have altered the life schedules of salmon and other commercially important fish. Pacific salmon, for example, may be responding in much the same way as the guppies, spawning at smaller sizes and younger ages than in previous years. The response is so widespread that it has been called "fishery-induced evolution," and scientists are hard at work studying its effects.[25]

Not all of the fishing-related changes mean smaller individuals, and not all are limited to the creatures being fished. Matt Wolak and other researchers in the southeastern United States studied diamondback terrapins, beautifully patterned turtles that are kept as pets, but in their natural habitat live in brackish waters along the Atlantic and Gulf Coasts. No one fishes deliberately for the turtles, but the youngsters sometimes end up in commercial crab traps and drown. Like many other kinds of turtles, the diamondback females are larger than the males, and after a terrapin becomes large enough, it cannot fit into the traps. The females can thus be safe once they mature, but the males almost never grow that big, so they remain vulnerable.

The researchers compared a Chesapeake Bay population of the terrapins (where the mortality of turtles in crab traps is high) with a population of living terrapins from Long Island Sound and with preserved Chesapeake Bay museum specimens from a period before the commercial crabbing had begun. The males from the areas with the traps were much younger, and the females grew faster and were 15 percent larger than their counterparts from the areas without the traps.[26] Again, these changes occurred within just a few decades and underscore the ability of animals to evolve in a variety of ways; they can become bigger or smaller and grow faster or more slowly, depending on the type of selection that is imposed.

The terrapins illustrate how rapid change can occur as a side effect rather than a direct consequence of hunting or fishing of the animal or plant in question. Similar side effect evolution has recently been seen in a fish, the Atlantic tomcod, which has evolved resistance to pollution released into the Hudson River.[27] Two General Electric plants released about 1.3 million pounds of polychlorinated biphenyls (PCBs) into the river between 1947 and 1976. The tomcod, which feeds on tiny shrimp and other invertebrates in the silt of the river bottom, did not react well to the chemicals, exhibiting, as Isaac Wirgin of New York University and his colleagues say with masterful understatement, "increased prevalence of malformations that were incompatible with survival."[28]

After just a few decades, though, the tomcod began to recover, and in 2011 Wirgin and coworkers published a paper detailing not only the recovery itself, but the exact genetic mechanism behind the tolerance of the pollutant. A tiny change in a single gene meant that the pollutant was metabolized differently while the fish were developing.[29] The discovery does not mean that we can expect all species to respond to pollution or other environmental threats with a flick of the genetic switch, of course, but it does suggest that it is possible for at least some organisms to evolve a way to survive despite damage to the ecosystem.

What about larger-scale changes? Fifty years on, even with their newfound ability to swat away the effects of PCBs, the tomcod is still unmistakably a tomcod, and the bighorn sheep are still bighorn sheep despite their more modest horns. Can rapid evolution even result in new species? The answer is yes, at least if you stretch the definition just a bit. Kathryn Elmer and colleagues at the University of Konstanz in Germany studied cichlids—not the kind that prey on guppies, but some of their relatives that live in the lakes formed by the craters of extinct volcanoes.[30] Cichlids are a very diverse group of fish, with numerous species occurring in many parts of the world, often coexisting in the same body of water. It has long been suggested that the different species arose quite rapidly, where "rapidly" means several hundred generations—not rapid by the guppy standards, but astonishingly fast compared with many other speciation events in the fossil record.

Elmer and her colleagues focused on a very young crater lake in Nicaragua: Lake Apoyeque, which formed about 1,800 years ago. This lake has two distinct forms of a kind of Midas cichlid. One form looks like an average fish, but the other has fleshy, protruding lips that make it look like the Angelina Jolie of cichlids. The exaggerated puckers are thought to help the fish wiggle invertebrate prey such as worms out of cracks in rocks, and are present in about 20 percent of the cichlid population in the lake.

When the scientists measured the head shape, diet, and genetic differences between the thick-lipped and thin-lipped cichlids, they found that the two forms were quite distinct. What's more, the fish seemed to mate mostly among their own kind, with the pouty males seeking out similarly plump-lipped females and vice versa. This separation means that the two forms are heading toward becoming different species in the same lake, all in less than 2,000 years—a very tiny interval for such a major evolutionary event.

A warming world of evolution?

What all these rapid evolutionary changes—tomcod able to resist pollutants, finches with bigger or smaller beaks—have in common is a dramatic and often sudden change to the environment, whether that change is the introduction of PCBs, a drought, or a group of hunters determined to bring home a trophy. The event causes some members of the population to have a much greater advantage at reproducing, like the salmon that mature at a smaller size when fishermen take most of the large ones in their nets. As the finches in the Galápagos show, the event does not necessarily have to be caused by humans, but human-induced rapid evolution is common simply because people often act at large scales: we don't remove just a few salmon, we take thousands of pounds of them. Commercial fishing is a gigantic selection force, and strong selection means that the differential survival of some genes over others, the essence of evolution, happens more quickly. When species are introduced to a new locale—carried either unwittingly in cargo or suitcases, or deliberately, as with the cane toads—they become more likely to evolve rapidly, because the new environment provides a host of selection pressures that force the newcomers either to adapt, as the toads did, or to die out.

Many species can survive quite well alongside humans, but even when they do, they may evolve new characteristics. We all know that a noisy highway makes it hard to sleep, but in addition, animals have a difficult time hearing each other when traffic and other noises of civilization are nearby. Hans Slabbekoorn at Leiden University in the Netherlands found that traffic noise means that songbirds such as European blackbirds and great tits don't communicate as well. They also have fewer offspring in noisy environments. In some cases, the birds shift the pitch of their songs so that they are more audible in a world filled with honking trucks and screeching brakes, but not all species are able to accommodate such intrusions.[31] We don't know whether all of the changes that Slabbekoorn and his colleagues have documented represent genetic changes in the birds, and hence rapid evolution, but the opportunity for human-induced evolution is clear.

Perhaps the most recent major environmental shift that has caused rapid evolution in many kinds of plants and animals is global climate change. The environments of many species around the world have been getting warmer, and quickly. In turn, birds are breeding earlier in many parts of Europe, and pitcher plant mosquitoes have shifted the time of year when they emerge from dormancy to become adults. Yukon red squirrels can have litters earlier in the season because the spruce cones they eat are ready sooner in the spring. Scientists are documenting many such changes in the timing of life events, as well as in the shape, size, and color of individuals.

Some of those changes are simply on-the-spot responses to the current conditions, rather than inherited changes passed on to offspring, but others are truly evolutionary in nature, permanently altering the genes of the individuals that make up the population. Still other species, of course, have not changed in the face of rising temperatures or melting glaciers, and face the risk of extinction. The potential effects of climate change on the world's animals and plants means that understanding rapid evolution, and the reasons why some species change while others do not, is more important than ever. At the same time, we should not use the evidence of rapid evolution to become complacent about the ability of animals to deal with human-caused disturbances; not every species is capable of shifting its schedules or acquiring larger beaks.

Finally, what about people? Where do humans fit on the list of species most likely to evolve quickly in the face of selection? It is ironic that we have induced rapid evolution in so many other species but, as I noted in the previous chapters, seem to doubt its occurrence in our own. Do we have what it takes? Criterion number one is strong natural selection, often via human-caused forces such as pollution or urbanization, but also through sieves such as epidemics, floods, and famines. We certainly qualify on those grounds.

Rapid evolvers also need the genes that will enable them to respond to such selection, and not all genes are equal in this regard. Biologists sometimes talk about "genetic architecture," a phrase I've always found rather elegant. It means the way that the DNA itself translates to what we see, with some traits, like the chirpless crickets, relying on a single gene that changes a pathway during adult development, and others requiring a group of different genes in different places to all work together. In Chapters 4–6 we will explore how selection, together with the buttresses and scaffolding of our own genes, leads to rapid evolution in people.

4

The Perfect Paleofantasy Diet

MILK

People are often passionate about dairy products. As I mentioned in the Introduction, my friend Oddný Ósk Sverrisdóttir and her colleagues are busy searching through the bones of long-dead Europeans for clues to the likelihood that milk was a part of their diet. And arguments about milk being a "natural" or even a healthy food abound. The website Notmilk.com, which, as the name suggests, is not a fan of the substance, intones, "Human milk is for human infants, dogs' milk is for pups, cows' milk is for calves, cats' milk is for kittens, and so forth. Clearly, this is the way nature intends it."[1] It also contains instructions on how to "detox" from dairy, promising that after just one week of abstention, "one gallon of mucus will be expelled from your kidneys, spleen, pancreas, and other internal organs,"[2] followed by an improvement in your sleep, mood, and sex life, though presumably the mucus needs to be dealt with before this rejuvenation can be realized.

This rather unappetizing image aside, and recognizing the considerable disagreement by medical experts with such an extreme view, it is undeniable that milk consumption is a novelty in human existence. Even today, the majority of people on Earth do not consume dairy foods after childhood. And that is just the way it should be, according to some readers of the *New York Times Well* blog and the website Cavemanforum.com. One of the former stated:

> Cheese is dairy. People aren't designed to eat dairy. Baby cows are. Simply put, it's a large reason people are obese and suffer western diseases like diabetes, heart disease, cancer and auto-immune diseases. There's a HUGE body of research behind the evils of cow proteins. Cow milk (and the cheese made from it) is designed to make baby cows fat quickly. Guess what it does to humans?[3]

On the forum, one response to the question "What's wrong with cheese?" was:

> I don't eat it because it's not paleo, as Paleo Dude says. The basic idea behind paleo is that we are adapted to what we ate during the 2,000,000 years of the paleolithic, so those things are safe. Anything else is questionable.[4]

It is precisely this novelty that makes milk, or more accurately, the ability to digest it, the poster child for rapid evolution in humans. We understand more about how this ability came about, what it means at a genetic level, and what its consequences have been, than we do about virtually any other change in the human genome.

What's more, the use of dairy as a food source is an illustration of how genes and human cultural activities can influence each other. This gene-culture interaction is also huge: it means not just that humans have evolved, and recently (impressive though that is) or that culture has changed through time (though that, too, occurs), but that both have been altered, one by the other, in a tight coevolutionary spiral that may be continuing right now.

Finally, understanding the evolution of lactose digestion requires the use of every tool that science possesses. We use computer models that treat people and their genes as bits of code in a hypothetical universe. We use the dusty particles of DNA that Sverrisdóttir and her coworkers gouge from bleached bones, as well as the juicier samples extracted from modern Finns and African nomads. We even use microscopic bits of butterfat scraped from shards of ceramic vessels that were used to store milk and cheese thousands of years ago. Who knows—maybe someone millennia from now will be delving into the discarded Häagen-Dazs cartons of our time to find out more about our genes.

The problem

Unlike some animals that specialize in only one or a few foods—for example, anteaters that eat just their eponymous insects or koalas that monotonously munch eucalyptus—human beings can eat and digest a wide variety of foods, both plant and animal. We cannot eat everything, of course; we lack the bacteria and other microorganisms necessary to break down the cellulose in grass and many other plants, which means that we cannot survive by grazing like cattle. But many other foods are fair game.

At first glance, milk would seem to pose little difficulty for omnivores like us. After all, milk, or the ability to produce it, is what defines us as mammals, along with mice, whales, goats, armadillos, and the rest of the variously furred or haired crew. Our group got its name back in 1758, when Carolus Linnaeus, the Swede famous for developing the double-barreled system of Latin names for each organism still in use today, decided to make "the female mammae the icon of that class," as historian Londa Schiebinger puts it.[5] Schiebinger points out the gender politics evident in Linnaeus's choice, observing that no other group was distinguished by female reproductive parts, but be that as it may, milk is essential to the young mammal's life. Lactation, the production of milk, provides mammalian offspring with nutrition and essential components of the immune system. Even the monotremes, those oddball mammalian egg layers like the echidna and platypus, rely on milk, with the youngsters lapping it up from specialized skin on their mothers' bellies. We could, with equal justification, have been dubbed "lactimals."

All good things must come to an end, however, and so it is with milk consumption, or at least so it is for all mammals other than humans. Each species has its own signature blend of components such as protein, fat, and calories geared to the growth schedule of the young animal consuming it. Cow's milk is higher in protein but lower in fat than human milk, though it contains nearly the same number of calories. Whales and seals are famous for the high fat content of their milk, essential to the rapid growth of their young in cold environments. A colleague of mine who studied seals in the Arctic says that the milk in some species is so thick with butterfat that it resembles toothpaste squeezed out of a tube into the mouths of the hungry pups. Surprisingly, mouse milk is quite high in fat and calories, though not on a par with the milk of marine mammals; a cup of mouse milk, assuming you had the patience to obtain it, would contain over 400 calories, more than two and a half times the number in cow's milk.[6]

Despite these differences, nearly all milk, regardless of the species from which it originates, contains a type of sugar called lactose. Digesting lactose—that is, breaking it down in the small intestine so that it can be used by the body—requires an enzyme called lactase. The ability to produce lactase is genetically controlled, and that ability is present, with extremely rare exceptions, in all mammals at birth. The lactase molecule spends its time tucked into the small intestine, able to encounter lactose as it tumbles by from dairy-rich meals. (A small percentage of babies are allergic to milk at birth, but this allergy is a reaction to the protein in milk, not its sugar, and it is distinct from lactose intolerance.) A funny thing happens on the way to adolescence, however: lactase production in all non-human mammals, and in most humans as well, grinds to a near halt sometime soon after weaning. Most adults possess only about 10 percent of an infant's lactose-digesting ability. Why lactase stops being produced is an interesting question that has received, at least in my opinion, too little consideration. Saying "it's not needed anymore" isn't really satisfying, since of course organisms retain many characteristics that aren't necessary, from human appendices to the tiny vestigial legs in whale skeletons.

Whatever the reason for lactase's usual disappearance, consuming lactose after the enzyme is no longer active often has unpleasant gastrointestinal consequences. If it is not broken down, lactose passes into the large intestine, which contains the rich stew of microorganisms we rely on for help with digesting many of our foods. When these microbes encounter lactose, they cheerfully ferment it too, causing the production of methane and hydrogen gas. We always produce some of this gas, but in large amounts in the lower gastrointestinal tract, it, along with other by-products of the bacterial activity, causes bloating, abdominal cramps, and diarrhea.

Some drugs and parasites, such as the protozoan *Giardia*, may damage the intestine so that it no longer digests lactose, but the more usual cause of lactose intolerance is the absence of the enzyme. A common test for lactose intolerance measures the amount of hydrogen gas exhaled in the breath after consumption of milk products; in people who don't produce lactase, the levels are higher. Even those without lactase can often consume small amounts of lactose, and fermented dairy products like cheese and yogurt are usually tolerated reasonably well, since the lactose has already been partially digested in such foods. But generally speaking, if you lack lactase, dairy products are not a desirable part of the menu.

The solution

Despite these difficulties, many people can continue to consume dairy products throughout their lives. Why is that? The Greeks and Romans noticed that adults differ in their ability to digest dairy products, but not until the last half of the twentieth century did people recognize patterns of lactose tolerance that suggested it was genetically based. In people who can tolerate milk as adults, the gene responsible for producing lactase continues to be active because of a mutation in another genetic region that ordinarily curtails the enzyme. By the 1970s, scientists had determined that lactase persistence is a dominant trait, meaning that only one copy of the gene controlling it, from just one parent, is needed for a child to exhibit the ability to digest dairy. (If it were a recessive trait, in contrast, a child would need to inherit two copies of the gene.) The exact nature of the molecules governing lactase persistence was identified in the early part of the twenty-first century, although work on the details of the alteration continues.

Initially, lactase persistence was assumed to be the "normal" state, with the lactose-intolerant an unfortunately flatulent few. But as more people around the world were surveyed, it became clear that the ability to digest dairy products in adulthood is prevalent in only certain parts of the world: northern Europe, particularly Scandinavia, and parts of Africa and the Middle East. About 35 percent of all the people in the world show lactase persistence, and they are clustered in only a few places.

The distribution in Africa is particularly interesting, because there, ethnic groups that live virtually side by side can have strikingly different proportions of individuals with lactase persistence. In the Sudan, for example, lactase persistence is seen in less than 30 percent of the Nilotic peoples but occurs in over 80 percent of the nomadic Beja. The Bedouin of the Middle East are also much more likely to be able to consume dairy than are their non-Bedouin counterparts in the same regions. What is responsible for the variation? Rapid evolution.

Cows first

To understand the evolution of lactase persistence, and the reason for the odd distribution of the genes that allow it, you first need to think about cows. And not just about the animals themselves, but about the relationship between humans and cattle or other milk-producing hoofed animals—a relationship that extends back at least 7,000 years.

Cattle were originally domesticated for the usefulness of their meat and hides, not their milk, and some cattle-keeping humans still make little or no use of dairy products in their diets, relying on the animals for meat and hides. If you want to obtain milk from cows for human consumption, you have to take the calves away from their mothers while the mothers are still producing milk, and then selectively breed the females that are best able to keep up that production, eventually creating cattle that differ from their ancestors in the genes that control lactation.

To even think of making use of cow's milk, though, you have to have cows to begin with. And not all early peoples raised cattle. So what gave people the idea to raise cattle in the first place? Or, to put it another way, as Gabrielle Bloom and Paul Sherman of Cornell University asked, what kinds of environments make it too difficult to keep cattle or other ungulates year-round?[7]

Bloom and Sherman figured that in extremely hot or cold climates, or places where little food for the herds was found year-round, cattle would be too difficult to keep. More important, the prevalence of animal diseases such as anthrax and rinderpest in certain parts of the world can make it virtually impossible to successfully rear cattle and keep them healthy. Bloom and Sherman examined the relationship between where we see ancestral lactase persistence in the world and the geographic distribution of nine of these diseases, as well as the climate in which people with lactase persistence originated.

As they had predicted, people from parts of the world where the diseases had been prevalent were far less likely to exhibit lactase persistence, as were groups from extreme climate regimes, such as deserts or tropical rain forests. A few interesting anomalies emerged from this analysis: some groups of people live in areas that should be inhospitable to cattle, such as near-equatorial Africa or the Middle East, but nonetheless keep dairy herds and have a reasonably high rate of lactase persistence. Bloom and Sherman speculated that these people were able to overcome the barrier posed by the diseases by being nomadic, moving themselves and their cows around to avoid infection.

The cattle of today possess genes that are very different from those of their ancestors. Some of the change is a simple directional one; most modern domesticated dairy cattle produce far more milk over a much longer period, for example, than do either modern beef cattle or the forebears of either type. If you want more milk, you encourage early weaning; cows reared for meat are weaned late because the calves grow larger that way. Careful analysis of the teeth from cattle skulls dating back to the Neolithic reveals that the calves in such early herds were indeed taken from their mothers at a relatively young age, supporting the notion that cows were first domesticated for milk. In addition, where cattle are commonly kept, as in north-central Europe, the genes for different components of milk, such as the specific proteins, are quite variable. According to a research team led by Albano Beja-Pereira of France, this diversity of genetic types reflects the selective breeding of cows by early pastoralists, rather than new mutations arising in the different populations.[8]

A chicken-or-egg question, without chickens or eggs

The earliest remains of domesticated cattle date back about 8,000–9,000 years. Initially, before the controlled breeding mentioned in the previous section, and even now in some parts of Africa, the animals were used not as a source of fresh milk but for pulling plows or as sources of meat or blood. Fermented milk in the form of cheese or yogurt was sometimes consumed, but these dairy products usually have little lactose and hence cause fewer digestive problems than fresh milk does.

To understand the evolution of lactase persistence, it's important to know which came first: milk drinking or the change in the lactase gene. In other words, did early humans somehow evolve the ability to tolerate lactose, via the random changes in genes that occur all the time, and then start to use dairy and fine-tune their domestication of cattle; or did the gene for lactase persistence evolve because of the evolutionary advantage of dairy consumption? Finding the answer has required some of the most sophisticated genetics, and archaeology, of our time.

When the relative proportion of a gene in a population changes through chance, the process is called genetic drift. Drift happens in most if not all populations, but it is most common in small groups, simply because a change that is neutral with respect to evolution—providing its bearer with neither benefit nor cost—is more likely to become fixed in a population if there are not a lot of other options.

The concept of genetic drift is relevant to our understanding of lactase persistence because it is a kind of default hypothesis: instead of selection producing the change in gene frequencies, the revised proportion of milk digesters could have arisen by chance alone. Many forms of many genes come and go, and it is certainly within the realm of possibility that lactase persistence genes arose in a few human populations by chance—the actual chemical alterations required are not that complicated. The genes would then have become established in the population slowly, with nearby genes on the same chromosome as the lactase persistence genes changing independently of the ones conferring lactase persistence. In this scenario, one wouldn't expect to see big blocks of the same genes continuing to be associated with the lactase persistence genes.

Alternatively, if the gene rose in frequency not by chance, but because of selection favoring individuals who could digest milk, a different genetic pattern would be expected. In this case the genes surrounding the lactase persistence gene should be relatively more homogeneous than would be expected by chance, because strong selection would be carrying them along with the lactase persistence gene.

To picture this situation more clearly, imagine that the different forms of the genes are beads on a string, and then further picture the beads being able to slide on and off, with the probability of sliding increasing along with the amount of time that the beads stay together. If a string with the bead representing the lactase persistence gene keeps being plucked from the rest, it carries away, willy-nilly, the genes located next to it. The lactase persistence bead is propagated because it is advantageous, and the beads alongside are simply "hitchhikers," to use the genetic terminology.

Comparing these predictions—similarity of the blocks of genes near the lactase persistence gene, versus random assortment—requires fairly complicated statistical analysis, but with the ability to determine genetic sequences at a fine scale, it was possible to perform the critical test during the first few years of the twenty-first century. The answer? Selection. The beads—genes—associated with lactase persistence were much more similar than you would expect under a leisurely drift hypothesis. People able to drink milk without gastrointestinal disturbance passed on their genes at a higher rate than did the lactose-intolerant, and the gene for lactase persistence spread quickly in Europe.

It may seem implausible that such a modest ability would produce such a large change in an entire population, but a gene doesn't have to raise the reproductive success of its bearer all that much to become established. Anthropologists Rob Boyd and Joan Silk calculated that as little as a 3 percent increase in the reproductive fitness of those with lactase persistence would result in the widespread distribution of such a gene after only 300–350 generations.[9] That's about 7,000 years—a blink of the evolutionary eye. Other calculations have estimated that the forms of the gene allowing milk consumption are anywhere from about 2,200 to 20,000 years old—again, a shockingly brief interval in our history. But it seems to have been enough for at least some of us to take a step away from our early human ancestors.

What this means is that our own genome has evolved because of our cultural practices. Tolerating milk led us to keep more dairy cattle, which in turn continued to favor the genes for lactase persistence. This kind of coevolution, while not unknown before, is a beautiful example of what anthropologist Pascale Gerbault classifies as "niche construction."[10] All organisms inhabit what ecologists call a niche, a world defined by the requirements of the species; frogs need water, flies to eat, and weeds to hide in, while mosquitoes require hosts with blood and a refuge from swatters. The human niche is similarly constrained, and one of its proscriptions used to be milk. But once the early pastoralists evolved lactase persistence, we could shape our own destiny, in at least that one modest attribute. As anthropologist Alan Rogers put it, "We live in a radically changed environment, that we ourselves created."[11]

Calcium, food, and water

What makes drinking milk so beneficial? Why did selection favor those few with the ability to digest dairy after weaning? The most obvious reason is that dairy provides a supplementary source of nutrition, something that may be scarce for pastoralists. But scientists have suggested at least two other possibilities.

First, according to the calcium assimilation hypothesis, lactose tolerance is advantageous at high latitudes, where sunlight can be scarce and hence vitamin D levels low, because it allows more efficient uptake of calcium, much the way the vitamin itself does. Milk drinkers would thus be more likely to avoid the debilitating bone disease of rickets.

Alternatively, perhaps alongside either supplemental nutrition or calcium assimilation or both, is the notion that milk, above all, is a source of uncontaminated fluid, something that can be in short supply in the deserts of northern Africa. Compounding the danger of dehydration that the scarcity of water poses, one of the effects of lactose intolerance is diarrhea, which dehydrates the body even more. Members of the Beja, a nomadic people who herd camels and goats in the arid lands between the Nile and the Red Sea, drink about 3 liters of fresh milk each day during the lengthy dry season. If they could not digest lactase, the lives of these pastoralists would be difficult, if not impossible.

Gerbault and colleagues examined the distribution of lactase persistence across Europe and the Middle East. How, they asked, does milk drinking correlate with the amount of dairy herding by a given group of people, the archaeological evidence of the start of cattle domestication, and the changes that have occurred in other variable genes besides the ones for lactase persistence? The scientists conducted computer simulations in which they entered a variety of starting conditions for all of these variables and then asked whether one or more of the hypotheses for the advantage of milk drinking—better calcium absorption or the coevolution of people and cattle—explained the patterns we now see. It was difficult to separate the ideas, since both are compatible with the northern expansion of milk drinking in prehistory, but Gerbault's team concluded that the calcium hypothesis was a likely force in European, though not Middle Eastern or African, lactase persistence evolution.[12] To test the idea more thoroughly, genetic samples from people in other parts of the world, including Asia, where lactase persistence is relatively rare, would be needed.

A final hypothesis about the adaptive significance of lactose tolerance—or more accurately, intolerance—was proposed in the late 1990s by B. Anderson and C. Vullo of the Division of Pediatrics at the Santa Anna Hospital in Italy. They noted that malaria is a common disease in places where people are less likely to have lactase persistence. What's more, people who do not consume milk can suffer from low riboflavin levels, particularly if they are otherwise poorly nourished. The parasite that causes malaria doesn't grow as quickly in cells that lack flavin, so Anderson and Vullo speculated that people without lactase would be protected from the disease.[13] In this view, the trait that evolved in response to selection was the inability to digest milk, not the persistence of lactase. This seems an unlikely scenario, however, given that all other mammals lose the ability to digest lactose after weaning, making lactose intolerance a more likely ancestral state. Further research also revealed no connection between malaria and the genes for lactase persistence.

Africa, the milky continent

Africa is a funny place when it comes to milk. Lactose tolerance in northern Europe arose well after early humans had left Africa. If we accept the idea that selection on humans in Europe caused them to evolve a high degree of lactase persistence, then what are some groups of people in Africa, like the Beni-Amir pastoralists in Sudan, doing with a 64 percent frequency of lactase persistence? Isn't it just too much of a coincidence to assume that such widely separated people evolved the same mutation at more or less the same time?

Geneticist Sarah Tishkoff of the University of Pennsylvania discovered the answer to this question by driving along the rutted—sometimes nonexistent—roads of East Africa, asking people from forty-three ethnic groups to participate in a study.[14] She and her colleagues asked 470 volunteers to drink a solution of powdered lactose dissolved in water. The scientists then took blood samples at timed intervals to collect information about lactose digestion and a sample of the individual's DNA at the same time. This test for lactose tolerance is not as accurate as the ones given in a Western medical facility, but it was far more feasible to administer in the field.

Tishkoff and her coworkers looked at their subjects' DNA for genetic variations near the lactase gene itself, where the lactase persistence alterations had previously been discovered. They found a consistent pattern of a few of these variants in people who could tolerate lactose consumption, but the genes that had become altered were different from the ones previously discovered in northern Europeans. In other words, Africans had evolved lactase persistence independently of the Europeans, with the same trait—being able to digest milk—arising from different genes. In both places, however, the people who can tolerate lactose are, with a few exceptions, the ones whose ancestors kept cattle and other milk-yielding animals.

The African version of lactase persistence also arose more recently than the European version, given that people did not start keeping herd animals in southern Kenya and northern Tanzania until just over 3,000 years ago. Thus, lactose tolerance is one of the best examples not only of rapid evolution, but of convergent evolution, the independent evolution of a similar trait via different pathways. Other cases of convergent evolution include the wings of birds and bats, which have the same function but are anatomically quite distinct, and the sonar-like echolocation systems of some whales, bats, and shrews. The lactase example is more cryptic than these, since although no one would have suggested that wings arose in the common ancestor of birds and bats, the idea that humans evolved lactose tolerance only once seems on the surface to be more plausible than the reality of multiple occurrences.

Interestingly, about half of the Hadza people of Tanzania were found to have the lactase persistence gene—a hefty proportion, given that they are hunter-gatherers, not herders. Why did the Hadza evolve a trait they don't use? Tishkoff and coworkers speculate that the gene might be useful in a different context. The same enzyme that enables the splitting of the lactose molecule is also used to break down phlorizin, a bitter compound in some of the native plants of Tanzania. Could the lactase persistence gene also help with digestion of other substances? No one knows for sure, but the idea certainly bears further investigation.

Microbes, milk, and what's to come

Despite the strong evidence for recent evolution of lactase persistence in humans via natural selection, a number of questions remain. How exactly does the mutation in a gene associated with the lactase molecule manage to keep the ability to break up lactose in adults? And why does lactose tolerance stop at different times for different people? Finns generally have high levels of lactase persistence, but some of them lose the ability to digest lactose as teenagers, rather than either retaining it throughout life or losing it as small children. Addressing this problem means understanding how genes regulate each other at different times during the life span—a fundamental question in human biology.

At a more practical level, what does the evolution of lactase persistence mean for modern people who are concerned about dairy consumption? It certainly puts to rest the notion that because dairy "is not paleo," it is not an appropriate part of the human diet. One's ability to digest milk simply depends on one's genes, and those genes have changed, at least in those of us whose heritage is rooted in pastoralists. (Note, as I mentioned earlier, that milk allergy, which is an immune reaction to the proteins found in milk, is distinct from lactose intolerance; both are genetic, but they arise from completely different genes.) Of course, people choose not to consume many foods they are capable of digesting, but the story of lactase persistence is an object lesson in escape from paleofantasy.

The science, of course, continues. A different way of trying to understand the question of variability in lactose tolerance might lie in another anomalous African population. At least some Somali people consume plenty of fresh milk but lack the genes identified by Tishkoff and others that should allow them to do so—rather a reversal of the Hadza. Catherine Ingram and others at University College London suggest that gut flora in the Somali may break down the lactose, making it easier to digest, in much the same manner that milk products such as yogurt and cheese are made digestible when their lactase is broken down in fermentation. Many people who are lactose-intolerant have an easier time digesting these products than they do fresh milk.[15] It is tempting to speculate that a particular set of bacteria and other microbes were subject to selection among the Somali, which meant that they achieved the same end as the Europeans and other Africans with the lactase persistence gene, but through yet another means.

The analysis of our internal microbiome, the diverse array of microscopic organisms living on and inside of us, is one of the most exciting emerging sciences. It may help us understand how the Somali, the Hadza, and the Finns evolved convergent digestive abilities, despite their wide geographic separation. Consumption of dairy exquisitely illustrates the ongoing nature of evolution, in humans as in other living things. Our ancestors had different diets, and almost certainly different gut flora, than we have. We continue to evolve with our internal menagerie of microorganisms just as we did with our cattle, and they with us.

5

The Perfect Paleofantasy Diet

MEAT, GRAINS, AND COOKING

In late 2010, headlines and blogs were full of a new discovery about Paleolithic humans. Every major news outlet and many of the more obscure ones covered the finding, with the usual interviews of the study's authors, comments by experts not involved in the research, and speculations about its implications for modern urban life.

The discovery was not a new fossil, or a redrawing of the map of human migration out of Africa. It was not an analysis of ancient DNA, or even of the human form at all. Anna Revedin of the Italian Institute of Prehistory and Early History in Florence, along with her Italian and Czech colleagues, detected bits of starch grains from plants, including cattail-like root particles, on the grinding stones from 30,000-year-old archaeological sites in Italy, Russia, and the Czech Republic. The scientists concluded that our ancestors were making flour and mixing the ground-up plants with water to make what one member of the team, Laura Longo, called "a kind of pita," cooked on a stone heated in the fire.[1]

Several related recent discoveries went more directly to the way hominins not only prepared but consumed starch: the teeth. A team of researchers led by anthropologist Amanda Henry analyzed the plaque clinging to teeth of Neandertals, which seem to have survived, in those preflossing days, with the remnants of meals intact.[2] The plaque is easily distinguished from dirt or other contaminants on the teeth, and Henry and her colleagues found clear evidence of grass seeds, date palms, and a few other plants. What's more, the outer parts of the starch grains were gelatinized—a transformation that occurs only via heating, which means that Neandertals cooked their food. Interestingly, the various plants found in the ancient teeth are ripe at different times of year, suggesting further that the Neandertals may have returned to various sites to harvest the grains. Another study by Henry and colleagues found similar plant remains in the 2-million-year-old teeth of *Australopithecus sediba* from South Africa.[3] The two individuals examined also had remains of bark and wood in their teeth, as well as a wide variety of leaves and other softer plant material. This diet bears a strong resemblance to that of most living primates, but not to the supposedly carnivorous human ancestors cited by the paleo enthusiasts.

So early humans ate crackers. What's the big deal? Modern versions of cattails and the other plants found in the remains associated with the grinding tools, such as bur reed, are known to be nutritious; archaeology blogger Kris Hirst noted that a single hectare of cattails produces 8 tons of flour.[4] The ground roots could be stored and transported, making people less dependent on the seasonal availability of game animals. Why wouldn't we expect our ancestors to have taken advantage of this useful resource?

The answer is that a reliance on starch in the diet calls into question the various forms of the so-called paleo diet, which, as I mentioned earlier, uses our ancestors as models for the way we should be eating. If people are vociferous in their opinions about milk, they are positively fervent in their feelings about grains and other carbohydrates as suitable components of the diet. "Bread" and "pasta" seem to be fighting words for many of the proponents of a diet more like that of early humans. From the *New York Times Well* blog, for example, comes this quote: "Cereal grains like wheat, oats, barley, rye, maize and rice we started to consume between only 5,000 and 10,000 years ago and we are not used to it. Genetically we are still the hunter-gatherers from 190,000 years ago, adapted to meats, fruits and vegetables."[5] Eschewing grains often seems to be accompanied by an enthusiasm for carnivory, as this comment from Cavemanforum.com illustrates: "I personally can never get enough of pork belly (bacon without the crap). Yesterday, I had it with breakfast, lunch and dinner! At dinnertime, I had a pound of salmon steak, but I decided to fry up some pork belly afterwards. It's a good feeling."[6]

The first formalized suggestion that we should be relying on meat and modeling ourselves, at least foodwise, after our cave-dwelling ancestors came from *The Stone Age Diet*, a 1975 book by gastroenterologist Walter Voegtlin. Like its more recent successors, including *The Paleo Diet*,[7] *We Want to Live*,[8] and *The Paleo Solution: The Original Human Diet*,[9] *The Stone Age Diet* bemoans the modern Western diet of processed food high in starches. Voegtlin disapprovingly describes the average North American dinner of the time, in which "The average meat serving is modest and is often replaced with baked beans, macaroni and cheese, soy bean meat substitutes, or peanut butter. An array of vegetables is dwarfed by a mountain of mashed potatoes and gravy or a giant baked potato. Most families have bread or rolls with butter at dinner. Dessert is about the same as lunch [pie, cake, pudding, ice cream], only more of it."[10]

This way of eating, according to the supporters of various forms of the paleo diet, is responsible for our current obesity crisis, as well as the various "diseases of civilization," including type 2 diabetes, hypertension, and atherosclerosis. The problem, they claim, is that humans have not adapted to be able to safely consume grains, agriculture having arisen a mere 10,000 years ago, and hence reliance on carbohydrates takes us on an unhealthy and dangerously untrodden path from the diet we evolved to eat, which is meat based. The diets differ in their details, with various amounts and types of vegetables permitted, but they share an emphasis on meat and strictures against virtually all sweets and products using flour. A sample breakfast might include fruit and pork chops with herbal tea; snacks might consist of dried fish or meat and walnuts.

Voegtlin is dismissive of vegetarians, advocating a virtually completely carnivorous diet based on his (inaccurate, as I will explain) view of our evolutionary history. He declaims, "Did anybody ever tell you that your ancestors were exclusively flesh-eaters for at least two and possibly twenty *million* years? Were you aware that ancestral man first departed slightly from a strictly carnivorous diet a mere ten *thousand* years ago?"[11]

Voegtlin also warns against misapplying the evolutionary principles he promotes, with a story that one only hopes is apocryphal:

> The puerile syllogism that 1) man descended from apes; 2) apes eat coconuts; therefore 3) man should eat coconuts, impelled German August Engelhart [*sic*] to gather about him a group of disciples dedicated to eating nothing but coconuts. The community migrated to and became established on a South Pacific atoll. A fanatical disciplinarian, Engelhart decreed imprisonment and torture for those deviating in the slightest from the coconut diet. When the atoll was captured by the British during World War I only one survivor of the company was found—Engelhart himself—his legs swollen from starvation and his body a mass of putrid ulcers. He died shortly after being taken from the atoll with its abundant fish and shellfish population, which could have saved all the "cocovores" from protein malnutrition and death.[12]

Needless to say, current versions of the paleo diet do not assume that vegetarians are sticking to one food, coconuts or not. The fans of paleo also acknowledge that they are not necessarily emulating ancestral diets in every respect, but using them as a jumping-off point. They still, however, eschew plant-based diets. The paleo lifestyle has been embraced by many people, with a number of informal online discussion groups arising in which practitioners share advice and ask questions. Cookbooks with titles like *Paleo Comfort Foods: Homestyle Cooking in a Gluten-Free Kitchen* and *The Paleo Diet for Athletes: A Nutritional Formula for Peak Athletic Performance* cater to specialized applications of the paleo diet.

The level of dietary and other detail is astonishing. Depending on where you look, people are advised to use coconut oil instead of olive oil, olive oil instead of coconut oil, or butter instead of either; to restrict nuts, or to eat macadamias instead of almonds; not to worry about eating nearly 6 pounds of ground beef in a day, but to be concerned about the high sugar content of watermelon; and to go to sleep when it is dark and wake when it is light, avoiding the use of alarm clocks. (Application of the latter recommendation to those people living in Scandinavia in the summertime is not addressed.) One practitioner consulted with a paleo discussion board because of a concern that eating more carbs was making his/her nose rounder ("when i stick to meat fat and veggies, it is pointier . . ."[13]). Others extend the back-to-the-cave movement to clothing, wondering which type of natural fibers—wool, silk, linen, or cotton—might be the most appropriate to wear (synthetics, of course, are out), calling to mind an interesting image of cave-based silkworm farms. Also, I can't be completely sure, but apparently some followers are eating horse fat. Is this really the solution to our health problems?

Amber fields of . . . game?

At a recent conference I attended,[14] Loren Cordain, author of *The Paleo Diet* and a well-known proponent of eating fewer refined grains as the key to health, gave a presentation to a small group of scientists interested in evolution and medicine. He outlined in persuasive detail the digestive and other health consequences of eating certain foods for people who have an alteration in a particular immune system gene. Potential offenders ranged from green tomatoes and root beer (not too much of a problem to give up) to bread, rice, and potatoes (ouch). Many of us can eat these foods with impunity, but to those unlucky enough to bear the gene variant, the foods can eventually cause something called "leaky gut," which you don't have to know anything about to know is something you don't want.

I was greatly intrigued by this information (and am cautiously optimistic that my gut is impervious, at least for the time being), but one thing puzzled me. Why, I asked Cordain, has this inability to properly digest all these common foods persisted? Surely it should have been selected out of the population.

He was taken aback. The answer was obvious, he responded. The sensitivity had been occurring only since the advent of agriculture, so humans haven't had an opportunity to adapt yet. I frowned. "Plenty of time," I said.

"But it's only been ten thousand years," he said.

"Plenty of time," I repeated. Now it was his turn to frown. We never resolved our disagreement, but it points to a question that is at the crux of this book. Is the diet envisioned by Cordain and other paleo proponents really "the one and only diet that ideally fits our genetic makeup,"[15] as his book would have it? Is it true, as he claims, that "just 500 generations ago—and for 2.5 million years before that—every human on Earth ate this way"?[16] We know that the lactase gene, for example, evolved quickly, allowing humans who could not have consumed dairy products without discomfort to do so if they inherited the gene. What about the rest of our digestion-related genes?

I want to make it clear that I am not discussing the relative health merits of the paleo versus Atkins versus Mediterranean diets, or any other particular way of eating. I am also not concerned with the details of exactly how much animal protein in the diet is necessary to constitute a real paleo diet, or which micronutrients would have been present in which quantities of food in the period before agriculture. I also realize that not all paleo dieters are doing exactly the same thing, and that not all are really attempting to replicate ancient meals.

What's more, although I confess to a certain bemused fascination with the minutiae of, for example, how large a yam one can consume and still have the overall diet be considered paleo, I realize that virtually all diets—vegan, vegetarian, organic, and so on—have their fringe adherents, some of whom can become quite vehement. Many paleo eaters complain about being chided by virtuous vegetarians; one commenter on a blog who has not "knowingly consumed grain products for about 10 years, as well as processed sugary things" is "often lectured by people how humans can't survive without grains and must eat them (usually they are slightly overweight, puffy people who complain of various ailments)."[17] And certainly, eating mainly lean meat and few or no processed foods may well be healthy for at least some people, particularly when contrasted to a diet of Cheetos and Coca-Cola. The question is whether the various forms of the paleo diet really do replicate what our ancestors ate, and further, whether that should serve as a guide for us in the twenty-first century.

Nevertheless, the studies of the quality of paleo diets are at best mixed in their evaluations. In 2011, *U.S. News & World Report* had twenty-two experts, including physicians and professors of food science and nutrition at universities and medical schools, rank a variety of popular and relatively unknown diet plans, including various low-fat and low-carb options, as well as the paleo diet.[18] Then the experts "rated each diet in seven categories: how easy it is to follow [1], its ability to produce short-term [2] and long-term [3] weight loss, its nutritional completeness [4], its safety [5], and its potential for preventing and managing diabetes [6] and heart disease [7]. We also asked the panelists to let us know about aspects of each diet they particularly liked or disliked and to weigh in with tidbits of advice that someone considering a particular diet should know."[19]

The magazine defined the paleo diet as "the way we ate when we were hunting and gathering: animal protein and plants . . . if the cavemen didn't eat it, you shouldn't either. So long to refined sugar, dairy, legumes, and grains (this is pre-agricultural revolution); hello to meat, fish, poultry, fruits, and veggies."[20]

The results? The paleo diet came in dead last, at rank 20. "Experts took issue with the diet on every measure. Regardless of the goal—weight loss, heart health, or finding a diet that's easy to follow—most experts concluded that it would be better for dieters to look elsewhere. 'A true Paleo diet might be a great option: very lean, pure meats, lots of wild plants,' said one expert—quickly adding, however, that duplicating such a regimen in modern times would be difficult."[21] Modern foods, as I detail later in this chapter, are often much higher in fat and sugar than their wild counterparts—a difficulty acknowledged by at least some of the paleo enthusiasts.

A few studies have attempted to examine the effects of switching to a paleo diet on various health measures, including weight, of groups of subjects. Unfortunately, as with all diets, doing a carefully controlled examination of the paleo diet is difficult. One such attempt took twenty healthy volunteers and measured variables such as body weight, waist circumference, cholesterol levels, and blood glucose before and after the three weeks of the study.[22] Only fourteen of the subjects managed to complete the entire three weeks, and they lost an average of 5 pounds, with a 0.2 decrease in waist circumference (they were not overweight to begin with). Blood calcium levels fell by more than 50 percent—not surprising, since the subjects were not allowed to eat dairy and received no special instructions on how to obtain specific nutrients from the foods they were allowed.

The National Health Service of the United Kingdom analyzed the study, which was widely reported as confirming the health benefits of paleo eating, and was not impressed.[23] It pointed to the absence of a control group—a comparison group eating in a different way, to which the participants could be compared—making it difficult to interpret the findings. The NHS also noted that a dropout rate of 30 percent suggested that the diet might be difficult to follow for most people, which means that regardless of its virtues, a paleo diet might not be a panacea for the diseases of civilization.

The hunter-gatherer table—or, mole rats and starch

Cartoon images of early hunters living exclusively off the flesh of mammoths notwithstanding, Voegtlin's bold declaration that our ancestors consumed exclusively meat turns out to be untrue. Studies of fossil hominins suggest that their sturdy premolar teeth may have been used either to open seeds or to chew starchy underground tubers and bulbs. One rather unusual clue about ancestral diets comes not from fossil humans themselves, but from the fossils of mole rats, bullet-shaped rodents that live in underground colonies with elaborate tunnel systems and rely on the enlarged potato-like roots of wild plants for both food and water.

A mole rat away from tubers is a sad and hungry mole rat, so when groups of mole rat fossils are discovered, it means that these underground food sources must have been nearby. And indeed, mole rat fossils occur at the same sites as hominin remains much more often than would be expected by chance, suggesting that early humans also used these roots and other starchy foods. They may have used them mostly as "fallback foods" in times of scarcity, but even so, such foods would have allowed early humans to survive periods when hunting was simply not an option.

Finding out that early humans ate starchy foods is one piece of the puzzle. The other source of information about what our forebears' diet might have been like is the diet of people now living the way we think our ancestors did, as foragers and hunters. As I mentioned in Chapter 2, there are many caveats to using modern hunter-gatherers as surrogates for early humans, but keeping those in mind, we can learn, for example, that the different cultures using foods they catch and glean eat a huge variety of different things, both plant and animal. It should come as no surprise that this variety is tightly linked to the range of climates and environments in which humans live; more northern peoples eat more meat and less plant material, while coastal cultures eat fish.

Anthropologist Frank Marlowe painstakingly examined the eating patterns and environments of 478 groups of humans across the planet, from the southern tip of South America to northern Alaska, and from eastern Australia to Africa.[24] He points out that whether you rely primarily on plant as opposed to animal foods depends on where in the world you live; Old World foragers get more of their diet from gathering and less from hunting and fishing, with the converse true for those in the New World. Even among the latter, however, nearly a third of the diet comes from plants, putting to rest the notion of our carnivorous ancestors. Climate matters as well, with those living in warmer parts of the world closer to the equator obtaining a larger fraction of their food via gathering, as would be expected, given the difficulty of acquiring fruits or other plant products in a harsh climate.

One of the most interesting distinctions in Marlowe's survey is between cultures that use bows and arrows for hunting and those that do not. Bows enable the killing of animals at a much greater distance than do spears, as well as the bringing down of larger game. As a result, it is possible to increase meat consumption, which Marlowe speculates could have spurred the growth in the human population after bows had become established in early societies. He further notes that if we are to use modern hunter-gatherers as models for the earliest of humans, we should probably stick to those groups that lack the bow, because they are the most similar to our ancestors, which leaves us with indigenous Australians and Tasmanians. Bow use is a deceptively simple advance with the potential to substantially alter the food base of a society, and it is yet another example of a subtle way in which preagricultural humans varied from a uniform hunter-gatherer model.

Another implication of the importance that Marlowe attaches to bow hunting is that, rather than starting out as exclusively carnivorous and then adding starches and other plant material to the diet, ancient humans would have been able to increase the proportion of meat only after newer technology had come about, a mere 30,000 years ago. Other anthropologists concur that the amount of meat in the human diet grew as we diverged from our other primate ancestors. All of this means that, first, contrary to the claims of many paleo-diet proponents, the earliest humans did not have an exclusively meat-based diet that we are best adapted to eat; and second, our ancestors' diets clearly changed dramatically and repeatedly over the last tens, not to mention hundreds, of thousands of years, even before the advent of agriculture.

What people eat is also inextricably entwined with how people spend their time, and in particular with how men and women might differ in this regard. I discuss paleofantasy gender roles in Chapter 7, but it is worth mentioning that the relative amount of the food budget supplied by men versus women, related to the reliance on hunting versus gathering, is another attribute that varies across cultures. Also tied to the appearance of bow hunting was the ability of men to provide a larger proportion of a family's caloric intake, since they were killing larger animals. Marlowe speculates that before this technological development, men might have been contributing different non-meat-based foods, such as honey, with women providing the bulk of the food eaten by the group, as was the case for the indigenous Australians in his sample.

Katherine Milton is a primatologist and anthropologist at UC Berkeley who has studied and written extensively about the diets of modern apes and monkeys, as well as those of hunter-gatherers and ancient humans. She points out that "it is difficult to comment on 'the best diet' for modern humans because there have been and are so many different yet successful diets in our species."[25] Furthermore, "Because some hunter-gatherer societies obtained most of their dietary energy from wild animal fat and protein does not imply that this is the ideal diet for modern humans, nor does it imply that modern humans have genetic adaptations to such diets."[26] Milton also takes issue with the attempt by paleo-diet experts to calculate the ratios of nutrients such as protein or saturated fat in hunter-gatherer diets so that these can be emulated by followers of the paleo way of life, again citing the vast range of diets eaten by modern foragers and the lack of detailed information on foods eaten by early humans.

Even if we did know more about what our ancestors really ate, Milton argues that it wouldn't necessarily help, because those ancestors do not seem to have been particularly specially adapted to their foods, or as she says, "regardless of what Paleolithic hunter-gatherer societies were eating, there is little evidence to suggest that human nutritional requirements or human digestive physiology were significantly affected by such diets at *any* point in human evolution."[27] The notion that humans got to a point in evolutionary history when their bodies were somehow in sync with the environment, and that sometime later we went astray from those roots—whether because of the advent of agriculture, the invention of the bow and arrow, or the availability of the hamburger—reflects a misunderstanding of evolution. What we are able to eat and thrive on depends on our more than 30 million years of history as primates, not on a single arbitrarily more recent moment in time.

Milton hastens to point out, and I wholeheartedly concur, that this does not mean that many of the diseases of civilization are unrelated to diet, or that we should go further back and emulate the diets of gorillas or other modern apes rather than hunter-gatherers. If the proponents of paleo diets were correct, however, those hunter-gatherers with higher levels of protein in their diets should be less likely to suffer from modern maladies like obesity and hypertension than do cultures consuming more plants and starches, but no evidence exists to support such a claim. Such diseases are indeed absent in most foraging societies, but they are more uniformly absent than would be expected if a high meat intake were the preventive. The biggest problem with many Western diets seems to be their energy density—the vast number of calories contained in a Big Mac as opposed to an equal volume of wild fruits or game. The concentration of calories makes it far too easy to overconsume, even when it feels as if the amount of food eaten is not particularly large.

The relative proportion of carbohydrates in the diet does affect one aspect of our health: our teeth. But here the problem arose not with a switch to agriculture, but much later, when the Industrial Revolution led to increased consumption of refined sugars and processed foods. A 2012 conference at the National Evolutionary Synthesis Center in North Carolina examined the dental health of modern humans compared with that of either modern peoples eating more traditional diets, such as the Maya of Mexico, or our fossilized ancestors. People in industrialized societies not only have far more cavities than either of the other two groups—Australian aboriginals from the 1940s were described as having "beautiful teeth"—but their jaws are shaped differently, with malocclusion and overcrowding of the teeth.[28] Ancient people and those consuming more fibrous foods simply chew more, which changes the development of the jawbones and associated musculature.

The scientists who studied early dental health are quick to caution against a quick fix of a paleo diet for those seeking to avoid the dentist and orthodontist; Peter Ungar from the University of Arkansas echoes anthropologists in saying, "There was not a single oral environment to which our teeth and jaws evolved—there is no single caveman diet. Still, we need to acknowledge that our ancestors did not have their teeth bathed in milkshake."[29]

One potato, new potato

Even if we wanted to eat a more paleo-like diet, could we? Virtually all of the foods available to those of us not living foraging lifestyles are vastly different from the forms that our Paleolithic ancestors would have eaten. I am not talking about ice cream or Pop-Tarts, or even flour, but about the unprepared basics: meats, fruits, and vegetables.

Many of the paleo-diet fans promote eating wild game, or at least very lean cuts of meat—a reasonable recommendation, given the differences in the fat content of domesticated animals compared to their wild counterparts. For example, according to the Department of Animal Science at Texas A&M University, a 4-ounce serving of white-tailed deer meat has 2.2 grams of fat, while a similar-sized portion of extra-lean ground beef has 18.5 grams of fat.[30] Game birds such as pheasant or quail have about half the calories of commercially available beef and pork. Domesticated animals that are used for food have been selectively bred, of course, to grow quickly, resist disease, and be amenable to living in much larger groups than their wild ancestors would have tolerated—all qualities that may or may not be compatible with a goal of maximal human nutrition. This is not to suggest that modern meat is unhealthy per se, but merely that eating like a caveman is going to involve more than skipping everything except the butcher counter at the supermarket.

The produce aisle has its problems too. Virtually all commonly eaten fruits and vegetables are also the result of much selective breeding by generations of farmers, whether one is growing one's own or shopping at the supermarket, using heirloom varieties or not. The ancestral potato, for example, was a bitter, lumpy root a fraction of the size of the average Idaho baker. The wild progenitor of apples, recently traced to Kazakhstan in Central Asia, was described by Michael Pollan as "a mushy Brazil nut sheathed in leather" that would "veer off into a bitterness so profound that it makes the stomach rise even in recollection."[31] What we now know as corn started out about 9,000 years ago as a Mexican grass called teosinte, with a shape and size more reminiscent of a stalk of rice than of the fat yellow kernels on a cob. Several researchers who followed wild monkeys or apes and attempted to eat the same fruits and other plants that those animals eat reported similar off-putting results.

Of course, the news isn't all bad. Milton examined eighteen species of wild Panamanian fruits eaten by several monkeys, and found that they averaged 6.5 percent protein, compared with just over 5 percent protein for cultivated fruits from the grocery store.[32] A similar analysis of fruits eaten by chimpanzees in Africa showed a protein content of over 10 percent.[33] Wild fruits also are more apt to contain worms and other insect infestations, probably adding negligibly to the protein content but, according to Milton, potentially increasing vitamins, amino acids, and other micronutrients not usually available in the fruit itself. And some fruits are quite large, sweet, and juicy even in an unmodified form, particularly tropical varieties like relatives of the soursop.

The point is that most modern foods, whether processed or not, are a far cry from their wild ancestors, having been enormously modified by people seeking a more calorie-rich, more easily transportable, or simply tastier version of the original. Many of the paleo-diet enthusiasts realize this, and the discussion boards are full of debates on whether to therefore stick to less sweet fruits such as avocados (to which other posters respond by noting that early forms of avocado are basically a large pit surrounded by a thin layer of, well, avocado flesh, which doesn't have much substance to it). The reality is that we are not eating what our ancestors ate, perhaps because we do not want to, but also because we can't.

In addition to shunning grains and the foods made from them, paleo-diet followers are concerned about eating starchy vegetables and tubers, such as potatoes. Sweet potatoes seem marginally better than the rest, but differences of opinion abound. One blog contributor eschewed all forms of potatoes, concluding by saying, "To summarize, there's 3 problems with tubers: a) Poisonous substances; b) Carb load; c) Problems we are yet to discover."[34] Arguably, this list contains either two problems or an infinite number, depending on your interpretation of point (c), but the interesting issue from my perspective is the first: the existence of toxins in tubers and many other wild foods.

Potatoes are from the same plant family as deadly nightshade, and the leaves and fruits contain several poisonous compounds, including solanine, which can cause digestive illnesses and, in large amounts, coma and death. Eating the potato itself avoids these substances, and one would have to consume quite a bit of the leaves or other parts of the plant containing these poisons to suffer ill effects. Nevertheless, the possibility of getting sick from one's food source unless one knows exactly which part to eat raises an interesting question: How did people figure out that the ancestors of potatoes were edible to begin with? Some other plant foods, such as acorns, require substantial treatment before they become nontoxic; Native Americans leached the poisonous tannins from acorns with successive washes before grinding them into acorn meal.

How in the world did people figure out what to do to render such initially inedible, even dangerous, plant products into palatable food? One assumes that necessity was the mother of invention, but the details of discovery, and the missteps—perhaps deadly—along the way, remain unknown. George Armelagos, an anthropologist at Emory University, suggests that distinctive cuisines developed as a way of restricting the potential for eating poisonous or otherwise unsafe foods in the environment.[35] That, too, however, underscores the flexibility of ancestral eating. We did not have a single diet, regardless of the relative compositions of fruits, nuts, tubers, and meat, throughout the period between the evolution of the genus *Homo* and the development of agriculture. Instead, people in different places ate different foods, modified them to differing degrees, and thrived on the variety.

The pot and the spit

Although the ability to digest the wide variety of carbohydrates available is more complicated than lactase persistence, involving more biochemical and physiological processes, in at least one respect humans have also evolved adaptations to eating starches, and relatively recently at that. Humans in different places have been eating different diets for thousands of years. The Japanese, for example, have been consuming starches in the form of rice for thousands of years, while people of the far north, such as the Yakut of Siberia, rely on hunting and fishing for sustenance.

A recent study by George Perry and colleagues asked the deceptively simple question of whether these two groups of people, and others like them, had evolved any adaptations to their different diets. As your mother may have told you, chewing food before swallowing it is important for proper digestion, in part because the enzyme amylase in saliva helps to break down starch. Salivary amylase also remains in the stomach and intestine after the food is swallowed, aiding the further digestion of starch by additional amylase present in the pancreas. Salivary digestion of starch, and therefore the presence of amylase, is crucial if an individual is suffering from diarrhea, since food is insufficiently digested in the intestine during such illnesses.

Perry and his coworkers looked for genetic differences in amylase genes from human populations with relatively low and high historical starch consumption: two African hunter-gatherer groups, an African pastoralist group, and the Yakut for the former; and European Americans, Hadza hunter-gatherers (who rely on starchy tubers) from Tanzania, and Japanese for the latter.[36] The researchers acknowledge that the diets are not mutually exclusive in their ingredients, but the two categories still show substantial differences.

The scientists examined not the structure of the amylase gene itself, but the number of copies of it in the genomes of the two sets of populations. Like many genes, the amylase gene is prone to duplication, with multiple copies existing in some individuals but not others. A person inherits the number of copies from his or her parents—in other words, the duplication does not occur during each individual's lifetime—but the likelihood that a duplication event will spread depends on whether or not it is advantageous. Gene duplication may have little or no effect, depending on what the gene does, but in the case of amylase, having more copies means that the individual is better able to digest starchy foods—an obvious advantage for people whose diet includes them. Perry predicted that the high-starch-consuming people would possess more copies of the amylase gene than the meat and fish eaters had. He was right; in the high-starch populations, 70 percent of the people had at least six copies of the amylase gene, but in the low-starch populations, that proportion dropped to just 37 percent.

Confirming the idea that the humans eating more starch evolved their higher number of copies of the amylase gene is the discovery that the levels of salivary amylase in chimpanzees are one-sixth to one-eighth what they are in humans. Chimpanzees eat a very low-starch diet, and geneticists Etienne Patin and Lluís Quintana-Murci from the Pasteur Institute in Paris suggested that a low copy number would have been found in our common ancestor, as well as in early humans.[37] (A section of their paper is titled "Adjusting the Spit to the Pot," which I stole as the heading for this section.) Then natural selection would have favored individuals with more copies as the amount of starch in the diet increased, so that the early agriculturalists could take advantage of the new foods. Cereal crops such as barley or rice probably were not domesticated until people had evolved more efficient starch digestion.

Interestingly, although chimpanzees and their close relatives the bonobos do not produce significant amounts of amylase, the Old World monkeys called cercopiths, a group that includes macaques and mangabeys, do. Perry and his colleagues speculate that the monkeys might use the enzyme to help them digest starchy foods, such as unripe fruit, that they store in their cheek pouches—a habit that only this group of monkeys possesses.[38]

Although we cannot go back in time to test early humans for the presence of amylase gene copies, an opportunity might exist, at least in theory, to do the next best thing. Until very recently, a small group of people in the remote mountains of northern Thailand and western Laos, the Mlabri, lived a nomadic hunter-gatherer lifestyle. But their language and culture, as well as the similarity between their genes and those of related peoples in the region, suggest an intriguing history. Hiroki Oota and colleagues believe that the Mlabri used to be agricultural but "reverted," in their words, to foraging, perhaps because the founding group of people was simply too small to support planting and harvesting of crops.[39] Depending on how recently the shift occurred (the data suggest anywhere between 500 and 1,000 years ago), it is possible that the Mlabri have a higher-than-expected number of amylase gene copies—a holdover from their farming past and a departure from other hunter-gatherer groups.

Agriculture's bitter legacy

Which other human genes associated with diet may have changed since the development of agriculture? Lluís Quintana-Murci of the Pasteur Institute has also studied an enzyme called NAT2, which was first described because of its role in metabolizing drugs used in treating hypertension and tuberculosis, but which is also important in breaking down toxins from plants and cooked meat. The gene that codes for NAT2 comes in several forms, with one form more common among hunter-gatherers and the other more common in people whose ancestors were agriculturalists.

Quintana-Murci and his colleagues believe that the variation in NAT2 has to do with the availability of folate in the two types of populations.[40] Folate is the naturally occurring form of folic acid, a substance known to many people because of its role in preventing miscarriage and birth defects, particularly spina bifida. Pregnant women are often encouraged to take folic acid supplements to ensure that their diets have enough of the compound. Folate is found in leafy greens or liver, neither of which are likely to be eaten in large amounts by agricultural peoples relying on grains for much of the diet.

NAT2 is also used to break down folate in the body, which means that having a gene variant that increases the rate of folate processing is a disadvantage if folate supplies in the diet are low; it's like enlarging the holes of a sieve when the flow of water is decreasing. So it's possible that natural selection favored people with less active forms of NAT2 after agriculture arose. At the same time, if folate is plentiful, metabolizing it more effectively could have helped reduce birth defects and the loss of pregnancies in hunter-gatherers. Support for this idea was found by a French research team led by Audrey Sabbagh, who surveyed the NAT2 variants of over 14,000 people from 128 different populations with a variety of diets. As expected if NAT2 is helpful when folate is abundant but harmful when it is scarce, foraging people were far more likely to possess the active form of the gene.[41]

Similar trade-offs between the advantages and disadvantages of genes associated with the diet are seen in the way we taste our foods. Humans have a remarkable sensitivity to bitterness—a useful ability, given the numerous bitter-tasting toxic compounds in wild plants—and several genes are responsible for the variation among people in their sensitivity to bitter flavors. Ordinarily, being more sensitive to bitterness is better, because it enables the taster to reject a potentially poisonous food source before eating too much of it. Among some people of central Africa, however, a gene variant that makes the bearer less able to taste bitterness is extremely common, despite the use of cassava, a starchy (and somewhat bitter-tasting) root, in that region. When some compounds in cassava are broken down in the gut, they release cyanide, which is obviously toxic in large quantities. If enough protein is present in the diet, however, people can tolerate small amounts of cyanide, so it is not automatically deadly. The less discerning gene variant may persist because it also renders those who carry it more resistant to malaria, which is endemic in the region. The distribution of the less sensitive variant mirrors that of the malaria parasite in Africa. Interestingly, some bitter-tasting plants in the region also have antimalarial activity, suggesting that humans with less sensitivity to the plants' taste may be more willing to ingest them.

Dietary signatures

Although the method of analyzing a gene here and an enzyme there has its appeal, many researchers are now taking a broader approach to the question of recent diet-induced changes in our genes. Modern genetic tools allow scientists to survey the whole genome—our entire collection of genes—at once, or at least to compare large chunks of it among different populations with different diets or ancestry. These techniques can uncover more subtle genetic changes than those affecting lactase persistence or amylase—changes that are caused by tiny alterations in many genes acting together, rather than by a single dramatic shift.

The basic idea is simple. Genes occur together on the chromosomes like houses on a street, with some neighbors closer than others. Like other organisms, humans have chromosomes in pairs, with alternate forms of a gene on each member of the pair. When sperm and eggs are produced, they each have only one member of the pair, so that when they join to make a baby, the child has the normal number of chromosomes. But before the chromosome pairs unzip, they swap some of the genes at locations across from each other; to continue with the house analogy, it is as if one of the even-numbered houses on one side of the street were exchanged for one of the odd-numbered houses on the opposite side. This genetic recombination is why children produced by the same parent differ from one another, even though they each get half their DNA from their mother and half from their father. Usually genes are swapped in big chunks, with several sequences of DNA linked together. The closer any single piece of DNA is to another, the more likely the two will be recombined together, so—returning to the house analogy—houses on the same city block tend to stay together for longer than do houses on adjoining blocks.

Suppose that a new gene allowing better digestion of grains arises in a population. Such a gene would confer an advantage to its bearer and hence is expected to spread quickly, as the children of the cereal lover, and then their children, pass on the gene. As they do so, however, these individuals also inherit the DNA that was located on either side of the gene, which is less likely to have been broken up in the process of recombination, simply because strong selection on the grain-digesting gene has taken its neighbors along with it.

What this means is that scientists can scan the genomes of people living in different parts of the world who have historically eaten different diets, or the genomes of humans and our close relatives the apes, and look for signs of suspiciously large chunks of DNA that seem to have been passed along intact, as you would expect only if selection favored a gene and then its neighbors were carried along with it in the haste of evolution. After these chunks are identified, it's possible to determine what the genes within the chunks can do: Are they associated with metabolizing protein? Do they detoxify plant compounds? Finding a large number of these changes means that our genes have changed relatively rapidly, in response to the new selection pressures caused by shifting diets.

Using these and similar techniques, Anna Di Rienzo at the University of Chicago and her colleagues found that humans show different genes associated with metabolizing food and accommodating the environment depending on where they live, or at least on where their ancestors came from.[42] Polar people's genes, for example, reflect an ability to deal with cold stress, and foragers appear to be adapted to a wider variety of foods than are the more specialized agricultural peoples. Liver enzymes associated with detoxifying the compounds found in roots and tubers are more prevalent in the nonpolar regions, where hunting is not the main source of food, as are genes associated with digesting cereals.

This branch of science, while attracting a great deal of attention, is still in its infancy; though researchers can identify and sequence stretches of DNA, assigning those sequences to a given function is not at all straightforward. And when many genes all contribute a small amount to a particular dietary ability, unlike the situation for lactase persistence or starch digestion, it is extremely challenging to chase them all down. Nonetheless, a strong body of evidence points to many changes in our genome since humans spread across the globe and developed agriculture, making it difficult at best to point to a single way of eating to which we were, and remain, best suited.

Sushi and dietary companions

The very newest discoveries about our dietary evolution may lie not in our own genes, but in those of the microbes we carry with us—in our guts, on our skin, and in the soil and air around us. When we digest food, we do not do it by ourselves, and I am not referring to having the family sit around the dinner table. We've all heard that some gut bacteria aid digestion, and perhaps that we can influence such bacteria by eating fermented foods like yogurt. But the real story is much more dramatic. The human cells of our bodies are outnumbered by microorganisms by a ratio of about ten of them to every single cell of ours. The microbes are part of our immune system and our sense organs, as well as our digestive tracts, and we could not survive without them. Pieces of their DNA lie intermingled with our own.

This microbial companionship might even have directed our dietary evolution. A recent study led by Jan-Hendrik Hehemann and his colleagues suggests a previously unconsidered mechanism for changing our genes under different dietary regimes.[43] The Japanese eat rather more seaweed than do other groups, averaging just over 14 grams per person each day. Seaweeds in turn harbor microorganisms of their own, including some that can help break down some of the carbohydrates in the seaweed. Unlike larger and more complex organisms, microbes from rather different groups can transfer genes to each other; the two types don't have sex exactly, but undergo a process called horizontal gene transfer. By this means, the genes from the seaweed microbes appear to have been transferred to the gut bacteria of people who tend to consume a lot of seaweed. The seaweed microbe genes thus were found in the intestinal flora of Japanese but not North Americans. It is clear that the Japanese did not get the genes simply by eating sushi, because one of the Japanese study subjects was a still-nursing infant who was not consuming solid food at all.

Other hints that intestinal microbes could affect the ability to consume new foods come from the finding that some people who do not have the lactase persistence gene can digest milk products; it is possible that they acquire microbes that do the job for them. Such gene transfer, and the interaction between us and our invisible companions, poses an extraordinary new avenue for rapid adaptation to the environment. Microbiologist Jeffrey Gordon says, "The gut microbial community can be viewed as a metabolic organ—an organ within an organ . . . It's like bringing a set of utensils to a dinner party that the host does not have."[44] As our diets change, so does our internal menagerie, which in turn allows us to eat more and different kinds of foods. The caveman wouldn't just find our modern cuisine foreign; the microbes inside of us, were he able to see them, would be at least as strange.

I realize that most people, whether or not they find the idea of genes hopping between the seaweed's internal flora and our own of interest, really want to know what these findings about the recent evolution of our diets mean for what they eat. The answer is that no one, whether a low-carb enthusiast, a proponent of bacon fat, or a fan of organic food, can legitimately claim to have found the only "natural" diet for humans. We simply ate too many different foods in the past, and have adapted to too many new ones, to draw such a conclusion. If the genetic tools to characterize each individual's diet-related genes are developed, we might be able to examine the suitability of particular diets for certain people. But that day has not yet come.

6

Exercising the Paleofantasy

What we eat is obviously a defining feature, not just for humans but for other creatures; we refer to animals as carnivores or herbivores. But what do funny-looking, glove-like shoes have to do with humans having bigger brains, less body hair, and a greater ability to sweat than other mammals? According to at least some scientists, quite a lot. And according to at least some influential runners, less is more when it comes to footwear, with no or minimalist shoes said to result in a more natural gait and fewer injuries. What's more, the way we move—in particular, the way we run—may be one of the most commonplace ways that humans show their unique evolutionary heritage. But are we best suited to run marathon-like distances, short sprints, or neither? And whether better distance runners or sprinters, were we necessarily faster, stronger, and fitter back—way back—in the day?

Are we being "seduced to sit"?

The "new age cavemen" profiled in the *New York Times* strive to emulate our ancestors not only in diet but also in exercise regimens.[1]

Some diligently follow a fitness program called CrossFit, which combines strength and cardiovascular training, including gymnastics, in a less structured manner than other workout routines; the founding principles include "Routine is the enemy" and "Keep workouts short and intense."[2] Although not explicitly attempting to mimic ancestral conditions or a paleo lifestyle, CrossFit touts itself as useful in "combat, survival, many sports, and life,"[3] and it advocates against specializing in a particular sport or workout type. Typical routines contain pull-ups, push-ups, squats, and running, as well as more specialized exercises called "pistols" and "Samsons," with the intensity and load varied for people of different abilities.

Other proponents of a more natural manner of activity include Arthur De Vany, whose blog is called *Evolutionary Fitness*, and who, in his mid-seventies, has a startlingly fit appearance that many half his age would envy. De Vany, like so many of the paleo followers, "begins with the premise that our bodies and minds are adapted to an ancient environment that passed more than 10,000 years ago," and that "by understanding the hunter-gatherer adaptation and incorporating the activity and eating patterns of our ancestral lifeway . . . we can live a natural and healthy life."[4]

De Vany emphasizes exercises that he feels emulate "the activities that were essential to the emergence and evolution of the human species. High intensity, intermittent and brief training mixed with power walking and play is closer than aerobic exercise, high volume weight training, or sedentism to how our ancestors lived."[5] Only with such behavior can humans be healthy and content, he claims. De Vany urges us to "embrace randomness and variety within the context of structured repetitiveness," which to him means eschewing daily long-distance running, for example, in favor of irregularly performed high-intensity sessions of weight lifting.

This idea of "exercising like a hunter-gatherer" is echoed by Loren Cordain of paleo-diet fame. He and others suggest that modern-day humans should attempt to perform activities that had equivalents in the Stone Age.[6] They provide a table outlining such parallel exercises, including "carrying a stacking rock," which, for those of us not constructing rock walls, can be handily translated into "lifting weights," while "gathering plant foods" is sensibly equated with "weeding a garden." "Carrying a young child" is considered, not surprisingly, an enduring form of exercise then and now, although "splitting wood with an axe" is said to be the modern-day equivalent of "butchering a large animal," which seems to me to substitute one difficult-to-replicate activity with another, at least for those of us in more urban settings. Like De Vany, James O'Keefe and company recommend alternating strenuous days with resting, to more closely replicate a life in which a group hunt, for example, would be followed by feasting and dance.[7]

The viewpoint that our lives would be better off if we moved as people did many thousands of years ago actually contains two different ideas, one of which is virtually unarguable, while the other is rather more controversial. The unarguable part is that many human beings today are relatively inactive, and this lack of activity is terrible for our health. Research about the detrimental effects of sedentary life on our health is accumulating from many quarters. Searching online for "sedentary lifestyle health risks" yields hundreds of thousands of hits, with couch potatoes suffering higher rates of mortality from various sources, as well as exhibiting more depression, anxiety, and other mental ailments.

The problem is more complex than merely a lack of exercise—whether exercise means hunting a gazelle, digging tubers out of rocky soil, or putting in forty-five minutes on the elliptical trainer; it's that people are spending vast amounts of time sitting. According to researchers led by Neville Owen at the Cancer Prevention Research Centre in Brisbane, Australia, "*too much sitting* is distinct from *too little exercise*" (emphasis in the original).[8] Recent studies suggest that even for people who manage to obtain the recommended amount of daily exercise each day, the deleterious effects of being sedentary can include larger waist circumference, higher blood pressure, and problems with blood sugar levels. The television appears to be a particular villain in this regard; the more time in front of the tube, the more drastic these adverse health effects were. The Australian researchers dubbed the conundrum of people who work out vigorously for a short period each day, but sit much of the remainder of the time, the "Active Couch Potato" phenomenon.

Active Couch Potatoes, or at least those from the population of white adults of European descent who were the objects of study, suffer higher-than-expected rates of mortality, even when some other potentially confounding factors, like the type of work they do, are taken into account. A survey of Canadians found that those who had reported spending most of the day seated had a higher risk of dying than did those who were even slightly more active. Although self-reporting is always a bit problematic for drawing conclusions, it is unlikely that the correlation arose out of a spurious association of, say, people who were simply too ill to move around being more likely to die. Among the Australians from the study that gave the Active Couch Potato its name, just a one-hour increase in the time spent watching television was connected to an 11 percent increased risk in mortality from all causes, and an 18 percent increase in dying from cardiovascular disease.

According to James Levine, a physician and researcher with the Mayo Clinic in Minnesota, such a sedentary existence is responsible for much of "the pandemic of obesity," as he terms it, in many parts of the world.[9] As a solution, he and his colleagues recommend increasing what they call NEAT (non-exercise activity thermogenesis), a fancy term for all of the things we do that expend energy but aren't officially considered "exercise," such as mowing the lawn, using the stairs, and walking to the kitchen. Such small actions add up through the day, burning what amounts to a significant number of calories as time goes on. Levine points to studies showing that highly active people—not athletes, just people who move around as part of their daily lives—have triple the energy output of inactive people. In addition, a sedentary person's metabolism changes in ways that appear to favor weight gain and alter levels of blood components like triglycerides that are risk factors for heart disease. In a 2010 article, Levine champions a "science of sedentary behavior," noting that we understand relatively little about either the ways to measure sedentary behavior (whether the physiological consequences of television watching differ in important ways from those of sitting in a car, for example) or the best interventions to take to reduce our tendency to be couch potatoes, whether active or not.[10]

Levine also "wonders whether exercise is a modern surrogate for the hunter-gatherer or agriculturalist life style," and he points out that although a tendency toward obesity has a significant genetic basis, those genes cannot have changed all that drastically over the last century, when the obesity crisis has become apparent.[11] Even an enthusiastic proponent of rapid evolution in humans and other animals such as me has to agree—it took thousands of years for lactase persistence to spread, not decades, and many more genes are involved in obesity. So if obesity is bad for your health and is influenced by genes, why weren't those genes selected out of the human population eons ago?

The answer, according to Levine, is NEAT, or the absence thereof. Suppose that some people are programmed to be "NEAT conservers" and don't move around much in situations when food is scarce, which would be adaptive if it enabled them to hang on to their fat stores until food became available again. Others are "NEAT activators" and increase their activity during famine, perhaps to seek food outside the usual hunting range. Until about a century or a century and a half ago, nearly everyone had to move around to live, using their bodies to do everything from washing clothes to visiting friends, so even though food was more readily available than it had been in previous millennia, most people burned off enough calories to avoid becoming overweight. With industrialization and the reduced need to expend energy, Levine argues, the NEAT activators still found a way to move around, whether by pacing, painting the walls of their kitchen, or going to the gym. But the NEAT conservers felt no such need, and, well, the rest is an ever-plumping history, at least for the last 100 years. "Over the last century," Levine says, "environmental cues have been so overwhelming that ambulatory individuals have been seduced to sit."[12]

This notion echoes the "thrifty genotype" explanation for the increased prevalence of diabetes in modern humans that was first promoted by anthropologist James Neel. Diabetes has become one of the major diseases of our time, costing the United States more than 15 percent of its total health care budget. Though diabetes occurs in people from all ethnicities and walks of life, it has risen particularly quickly among those of non-European descent, particularly in urban areas. While diet and exercise play an important role in when and even whether diabetes manifests itself in any particular case, genetics also plays a role, with people from those non-European backgrounds being more likely to develop the disease.

Those people, according to the thrifty-genotype hypothesis, have genes that efficiently utilize sugar and store fat during relatively infrequent times of abundance, which also enable them to survive during famine. The problem arises when ample food is always available; those same thrifty genes then promote hyperglycemia and problems with insulin regulation. People of European descent are less likely to have such genes because the boom-and-bust periods of food availability that selected for thrifty genes happened several hundreds of years ago, and people with a tendency toward diabetes were selected out. Furthermore, the frequency of famines decreased gradually. Among Pacific islanders and Native Americans, by contrast, particularly in urban areas, populations went from subsistence farming to convenience stores in a single generation, giving them little opportunity to adapt to continual plenty.

The basic idea in both cases—inactivity-induced obesity and overabundant food–triggered diabetes—is that behaviors adaptive in our ancestral conditions have become detrimental. Note that I am not advocating an exercise paleofantasy here; we are still not exactly like our Pleistocene ancestors, and evolution is still occurring. Just as it makes sense to eat foods that are not intensely calorie-rich all the time without promoting a diet of venison and tubers, it is reasonable to suggest that we are adapted to more movement without having to assume that humans are entirely unchanged since the Pleistocene. And Levine doesn't advocate a slavish return to activities that mimic those of a hunter-gatherer; spending the weekend putting up shelves rather than watching televised sports is fine with him, without having to try and run down a rabbit for dinner. This approach makes more sense to me than the rather literal-minded proscriptions of De Vany and the other paleo devotees. I do suspect it is more likely that the variation Levine notes in NEAT patterns is continuous, rather than falling into two neatly labeled categories. Levine himself hints at this in some of his writings, but as he points out, we need to start looking at how people move in a different way before we can draw firm conclusions.

Levine does not see the same sharp distinction between pre- and postagricultural people that the paleo crowd does; to him, and to many of those concerned about the health consequences of being sedentary, the problem came much later in our history, when mechanized tools and modes of transportation made that seduction to sit ever easier. Working on a farm, or keeping house without a dishwasher, still uses more calories than does commuting to a cubicle, even if one is not hunting prey with a spear or piling up those rocks.

What's more, the differences between mostly active and mostly sluggish do not apply solely to then versus now; they are just as stark in modern agricultural denizens versus city dwellers. For example, a 2011 study comparing rural and urban people in Jamaica and the United States found that the rural Jamaicans walked or ran

60 percent more than those living in the city of Kingston. When the US sample was divided into "lean" and "obese" categories, the latter were found to be spending nearly ten hours per day sitting, compared with only seven and a half hours for the urban Jamaicans and five and a half for the rural Jamaicans.[13] (Movement was measured using an elaborate electronic system of underwear sensors, which in one case resulted in a Jamaican subject being "mistakenly arrested because of the wired nature" of the apparatus.[14] One can only imagine the difficulty in explaining the study to the police.) Clearly, moving around more does not require dwelling in a cave and dressing in hides.

Even if our genes themselves have not changed in the last century, our prolonged periods of sitting might change the way those genes function in the body. Frank Booth and his colleagues at the University of Missouri in Columbia examined the physiological consequences of inactivity in both humans and rodents, with the latter serving as good experimental models because of their willingness to exercise on running wheels in cages. They concur with the idea that humans evolved to deal with a feast-famine cycle.[15] When rodents are suddenly prevented from running, they begin to store fat, particularly visceral fat, and their insulin sensitivity, an indicator of diabetes risk, returns to its low pre-exercise levels in an astonishingly short time, merely days or weeks. Studies of athletes who stop training suggest similar patterns in humans.[16]

It is possible that the genes responsible for metabolizing glucose behave differently in the body depending on whether that body is active or sedentary. A couch potato body sends the wrong signals to the genes, which behave as if a famine were imminent, since inactivity is historically associated with not having any food out there to go dig up, hunt down, or gather. The changes in "gene expression" (the scientific term for the actual behavior of genes under a particular set of environmental circumstances) affect a range of body traits, from muscle size to the constitution of the blood vessels. Scientists feel that exercise deficiency plays an important role in the modern prevalence of many diseases, from hypertension and atherosclerosis to Alzheimer's. Being still potentially sends too many inappropriate messages—"A crisis nears! Husband your resources for the short term, even if it means your body pays for it later!"—to a host of physiological systems.

Fitness and the snake in the library

The idea that being sedentary is bad for our health is one thing. It's another to suggest, as De Vany, O'Keefe, Cordain, and others do, that our hunter-gatherer ancestry means that short-term and variable exercise is good, while sustained-endurance activity, such as that performed by marathon runners, is harmful. De Vany recommends power walking with weights, to simulate our ancestors lugging chunks of mastodon from the kill site back to camp. He also suggests that this routine is less likely to incur the "pounding and damage" of jogging, of which he takes a dim view. Setting exercise or weight goals is, he claims, counterproductive because they "do not relate to function and process."[17] De Vany also admiringly recounts a story about "Indian braves" who killed five bison by driving them into a pit reported to be over 10 feet deep, after which the hunters pulled the 2-ton animals out and butchered them; the article emphasizes the strength involved in hauling the bison whole out of the pit, though I personally would have thought that a sensible hunter, Stone Age or otherwise, would have thought to render them into manageable pieces before doing all that heavy lifting.

O'Keefe and his coauthors provide a list of "essential features of a hunter-gatherer fitness regimen," which includes recommendations for exercising outdoors, including sexual activity for cardiovascular health. They are somewhat less negative about running than De Vany, but still state, "Humans in the wild were almost never walking or running on solid flat rock for miles on end." They are somewhat more in favor of "relatively softer natural surfaces such as grass and dirt," and also emphasize interval training, weight training, and alternating exercise days with rest days.[18] The paleo online bulletin boards are rife with discussion of how "authentic" running really is, with shamefaced confessions by runners unwilling or unable to give up their sport. Some refer disparagingly to "chronic cardio," as though it were a shameful medical condition. The followers are often particularly uneasy about the high-carbohydrate diets favored by marathon aficionados.

Both O'Keefe and colleagues and De Vany advocate group activities, with De Vany going so far as to note that "the evolutionary basis of sport seems clear. For example, the number of players in most popular team sports today is about equal to the number of prime age males that would be alive in a typical Paleolithic band of hunter gatherers."[19] This latter point, of course, is most charitably described as highly speculative, because both the number of "prime age males" in a given hunter-gatherer society and the team composition in many sports vary widely. Presumably De Vany was not thinking of the venerable sport of lacrosse, historical versions of which featured between 100 and 1,000 players.

Nassim Taleb, economist and author of the best-selling book *The Black Swan*, is also a fan of evolutionarily based exercise. He emphasizes "randomness" in movement,[20] which blogger and Taleb fan Daniel Patrick Johnson describes as similar to "when cavemen would spend most of their time meandering around looking for food, and then every once in a while they'd have to sprint away from a prehistoric tiger or chisel out a tool from a heavy boulder."[21] They would not have been lifting those boulders, however; Taleb suggests that "our ancestors had to face most of the time very light stones to lift, mild stressors, and, once or twice a decade, encountered the need to lift a huge one."[22]

Most anthropologists would hesitate to characterize the foraging behavior of modern hunter-gatherers as "meandering around," and even De Vany proposes that foraging could have substituted for today's aerobic exercise. I also couldn't help wondering what these hypothetical early humans would have used to "chisel out" a tool in the first place, since they would have needed, well, a tool, presumably made from yet another rock, to do so—which leads to a sort of Paleolithic infinite regression. In addition, the more standard method of manufacturing stone tools by flaking relatively small rocks, while requiring a fair amount of skill, does not seem as if it would require much exertion. Taleb also feels that the motivation behind exercise is important, so, for example, the aforementioned sprints are better if one imagines being angry or frightened, as if one were fleeing a predator or chasing a rival; he frets, "The only thing currently missing from my life is the absence of panics, from, say, finding a gigantic snake in my library."[23]

Born to run: Hunting, sweating, and hairlessness

Whether or not one views a paucity of large reptiles in the house as a problem, the ideas of Taleb and the other authors noted in the previous discussion are in direct opposition to a growing number of scientists and exercise fans who believe not only that running long distances is healthy, but that humans evolved doing just that.

To see why that might be the case, take a look at the human skeleton, preferably next to that of a chimpanzee, our closest living relative. Unlike the chimp, we are, of course, bipedal (though the chimp can assume that posture for short periods), and our bodies bear many signs of adaptations to walking on two legs: knees that can straighten out, a relatively flat "platform" foot, a curved lower spine, and well-developed gluteal muscles. When we walk, each leg acts like a pendulum, with energy stored in the body's center of gravity rising up during a step and then shifting to the opposite leg.

Because of its hip bone and pelvic structure, the chimp's feet are farther apart, and when the chimp walks, its comparatively shorter legs don't allow the same swinging gait that we humans exhibit. Humans are unique in keeping our legs relatively straight while we walk, yielding a characteristic bobbing motion that is not apparent in the movement of other bipedal animals.

Just how and why humans evolved the ability to move on two legs rather than four is still the subject of debate, with theories ranging from freeing the upper limbs for tool use or finding food in trees to better heat loss on the savanna. But as evolutionary biologist David Carrier pointed out in an influential 1984 article, when they think about the evolution of human locomotion, people usually think about walking, and rarely consider that we can also use those legs to run. To his mind, "The feature that differentiates hominids from other primates is not large brain size, but the set of characters associated with erect bipedal posture and a striding gait."[24]

Humans are often considered to be poor runners, at least in comparison to quadrupeds like horses and dogs, and it is true that our sprinting abilities pale next to theirs. We can't run nearly as fast as other animals (a greyhound, for example, can run 40 miles per hour, and the champion cheetah attains speeds of 70–75 miles per hour, compared with our paltry sprint times of 20-something miles per hour). And we burn more calories doing it; our "cost of transport," a term from animal physiology that refers to the amount of oxygen consumed per unit of body mass for every unit of distance traveled, is up to twice that of other mammals our size. This means that human beings are not particularly efficient runners either.

And yet, if you could hold a marathon for all the world's beasts, in a Noah's Ark of long-distance races, somehow compensating for differences in body size, the human runners would be close to breaking the tape first, leaving all but a few other species in the dust, including all of our closest relatives, the primates. Wild dogs and hyenas can run for long periods, as can the migrating ungulates such as wildebeests, but they are the exceptions. Monkeys and apes just can't compete when it comes to so-called endurance running. Even horses, well-known for their running feats, can falter when racing against humans; the horses need to rest more often, so a human runner can win if the race continues for a sufficient distance, particularly over uneven surfaces. On a more practical note, people from various parts of the world, both historically and today, have been able to hunt animals such as antelopes or kangaroos, simply by running after them until the prey collapse from exhaustion.

Carrier saw a conundrum: Why would people have evolved to do something that in many ways they are so bad at? Barefoot-running aficionado Christopher McDougall turns the question around, saying, "I never thought to ask, 'Hang on—if running is bad for humans, why isn't it bad for every animal?'"[25] In other words, either humans are in fact adapted to running, or antelopes need Nikes.

The answer seems to be that humans are indeed poor sprinters, but that this lack is more than made up for by our skills at distance running. In particular, Carrier cites two problems faced by the long-distance runner: getting rid of enough of the heat produced by the body, and storing enough energy to fuel the run.[26] The latter can be partially ameliorated with diet, but the former is more problematic. If you are a warm-blooded creature, vigorous exercise means that your metabolism generates quite a bit of heat, as any sweaty athlete knows. This heat production is why exercise warms you up on a cold day, but it also poses a risk of hyperthermia, or heat exhaustion, when the excess cannot be released quickly enough. That flash-on-the-savanna cheetah stores heat as it runs, raising its body temperature to dangerous levels in a relatively short time. Once that temperature exceeds 40.5°C, the cheetah won't continue to run. If it did, it would sprint for only about one kilometer—interestingly enough, about the same distance that cheetahs pursue their prey.

Humans, of course, have far less stringent limitations. As anyone who has gone for a run with a dog knows, when humans get hot, we sweat, but when dogs get hot, they pant, along with most other mammals. (Horses and camels are among the exceptions.) Sweating and panting both work to cool the body by evaporating water from its surface, but sweating has two advantages that are germane to a running animal. First, panting uses only the surface of the mucous membranes such as those inside the mouth, while sweating can take place over a large area of the body. Second, panting interferes with breathing, while sweating does not—an important consideration during strenuous exercise. What's more, being mostly hairless means that humans can off-load heat relatively rapidly while moving. As McDougall says in his book *Born to Run*, "humans, with our millions of sweat glands, are the best air-cooled engine that evolution has ever put on the market."[27]

Human running has another quirk: it lacks an optimal speed—a single rate of movement at which the cost of transport is the lowest. Horses, for example, need to switch gaits from trot to gallop or back once they reach a particular speed, while humans can adjust how fast they run relatively seamlessly and have about the same cost of transport for each speed. What's more, humans can alter their breathing during running to accommodate those different speeds, with, say, two steps per breath or three, depending on the need to oxygenate the lungs.

All of these characteristics led Carrier to propose an idea that was later further developed by Dennis Bramble of the University of Utah and Daniel Lieberman from Harvard University: the Endurance Running Hypothesis. To set the stage, Bramble and Lieberman list a variety of features of humans that are not shared with our recent ancestors and that aid running but not walking. These range from skull and neck structures that stabilize the head during rapid movement, to foot bones, including a stable longitudinal foot arch. The human head moves independently from the shoulder girdle, which doesn't matter when one is walking but helps keep the body upright during running.[28]

The existence of these differences between our muscles and bones and those of our closest evolutionary ancestors is important. Suppose we had elastic-like tendons in our legs and feet, but chimpanzees and gorillas did too. In that case, it would be hard to make the argument that those structures evolved because of natural selection on humans in particular to be good at running, since an obvious alternative argument is that we have them because the ancestor of all apes and humans had them too. A trait that is unique to a species doesn't always mean that it evolved through selection to perform a particular function, but it's a hint in that direction.

On the right track?

How did early humans use their flexibility in running speed, good endurance, and a superior ability to shed heat? According to the Endurance Running Hypothesis, the point wasn't running Stone Age marathons but hunting prey. When they are first startled by a predator, most prey animals sprint away, relying on speed to escape. But they tire after a relatively short period, and if they have not shaken the predator, the hunter can start to close in. Furthermore, most prey animals start to overheat after they have been running for a while, and will attempt to rest. If the hunter can keep after the prey, and not overheat itself, eventually it can catch up to its quarry, not because it is faster, but because the prey animal has started to develop heat exhaustion. So, humans evolved to be long-distance runners as a way to get food, and many of the other adaptations, such as our superior sweating ability, followed. Bramble and Lieberman conclude their paper by suggesting that endurance running allowed humans to acquire the protein- and fat-rich diet that many anthropologists believe allowed the evolution of our big brains.[29]

The process, logically enough called persistence hunting, actually sounds rather diabolical. Carrier points out that if early humans had the same cost of transport at various speeds, then they could "have picked the speed least economical for a particular prey type. This would have forced the prey to run inefficiently, expediting its eventual fatigue."[30] Keeping after an animal and not allowing it to rest when its temperature rises too much is more important, and more feasible, than overtaking it. In other words, running fast is not enough; you have to also run smart.

Running smart includes more than running; if you aren't at the heels of your prey every instant, you need to be able to track it, and tracking is a skill that comes only with long experience. It involves not only tracing the signs of an animal's passage through the wilderness by detecting subtle changes in soil or vegetation, but predicting where the animal is likely to go and keeping it away from shelter or even the safety of its fellow group members. Such tracking ability, not to mention persistence hunting itself, is rare among modern people, but Louis Liebenberg, a scientist and tracking expert from South Africa, has documented it in the Kalahari bushmen.

Over a six-year period, Liebenberg worked with several hunters in central Botswana, participating in some hunts and assisting filmmakers on others. When they were chasing large antelopes such as kudus, three or four men would drink as much water as they could and then set out while the air temperature was between 39°C and 42°C (102°F–107°F). They would then "run up to the animal, which quickly flees, and track its footprints at a running pace. Meanwhile, the animal will have stopped to rest in the shade. The hunters must find the animal and chase it before it has rested long enough. This process is repeated until the animal is run to exhaustion."[31]

Not all the hunters who began the hunt completed it, though the tracking skills of the older members of the group proved essential to success even when they did not keep up with the swiftest members. Hunts lasted approximately two to five hours, with an average running speed of 6.3 kilometers (about 4 miles) per hour, though the different hunters varied their speed. As Liebenberg points out, although this doesn't sound very fast, the process involved carefully observing the animal's movements and planning the approach—not simply head-down, ground-pounding forward motion.

The success rate of such persistence hunting is difficult to gauge, since so few hunts were recorded, but Liebenberg suggests that it may be about 80 percent, which compares very favorably with other hunting methods, such as hunting with clubs and spears or using dogs. The latter technique is the one that seems to have yielded the highest amount of meat per unit of effort, though again the number of documented hunts was relatively small. Such a success rate is also considerably higher than that obtained by mammalian predators such as lions or raptorial birds, which miss far more prey than they actually capture; one survey of raptors found that the birds managed to catch a meal only one out of five times.

After becoming infatuated with tracking and its effects on other parts of his life, Liebenberg developed software called CyberTracker designed to help record information from the environment in a user-friendly way, on handheld devices that can be taken into the field. The system can be used to monitor endangered wildlife, and Liebenberg even suggests its application to tracking criminals in remote areas. In a 2006 interview, he says, "With tracks and signs, you have to create hypothetical, causal connections between them, because you didn't see what the animal did. You have to visualize what the animal did."[32]

The increase in tracking skills with age and experience fits well with another observation about endurance running in humans: it is one of the only sports in which practitioners can improve with age and still compete effectively, well past the usual "prime" age of twenty to twenty-five. (Swimmers are said to hit their peak even earlier, at about age twenty-one.) In *Born to Run*, McDougall recounts a conversation with Bramble in which they calculated the decline of speed for marathon completion by men after the peak at age twenty-seven; the conclusion was that a sixty-four-year-old could run as well as a nineteen-year-old.[33] Whether that is strictly true or not, many people continue to run marathon distances into their seventies, eighties, and even, for a few, their nineties. What's more, women are extremely good at endurance running; in 2011, Amber Miller gave birth shortly after completing the Chicago Marathon, which is at least evidence that marathons are not something that only elite athletes with no other physically compromising conditions can master. Women also exhibit an extremely slow decline in their abilities. While these observations are not proof for the Endurance Running Hypothesis, they are certainly consistent with it.

To our prehistoric forebears, endurance running may also have been useful for obtaining food in another way besides hunting: scavenging carcasses that had either died of natural causes or been killed by other predators. Although the degree to which early hominins relied on such opportunities is debated by anthropologists, being able to outcompete either rival groups of humans or animals such as hyenas on their way to a fresh kill would obviously have been advantageous. Bramble and Lieberman point out that hyenas and wild dogs seem to detect their prey by smelling it or noting the circling of vultures overhead.[34] While humans might have relatively poor senses of smell, the ability to reach a distant carcass quickly after seeing vultures at a distance could have given our ancestors an edge.

If humans evolved as long-distance runners, then, why do most of us moan at the suggestion of even walking a mile or two? Despite the popularity of running as exercise and the recent upsurge in marathon participation, few of us regularly log scores of miles a week. Certainly some people enjoy it; in *Born to Run*, McDougall rhapsodizes about the joy and ease of running long distances, particularly without shoes, about which I will have more to say later in the chapter.

McDougall talks about running the way foodies talk about a new source of heirloom tomatoes, or wine enthusiasts extol their favorite vintage. He is not really a fan of formally organized marathons, preferring trail runs with "no fees, no awards, no whining."[35] Yet even he acknowledges that the appeal is far from universal. The answer to the paradox, he and Bramble suggest, is that in addition to an ability to run far and well, we have urges that tell us to conserve our energy—urges that functioned beautifully when a certain amount of activity was inescapable. Reminiscent of Levine's ideas about NEAT, the problem now is that we can easily give in to those urges, since we never have to go after that gazelle or compete with another tribe for a half-rotten carcass.

Our evolutionary exercise heritage may even suggest a way to fend off psychiatric disorders and Alzheimer's disease. A protein called BDNF (brain-derived neurotrophic factor) is essential to proper functioning of the prefrontal cortex, a brain region that also is impaired with mental illness. The protein increases during exercise, and scientists Timothy Noakes and Michael Spedding suggest that if our ancestors had evolved as endurance runners, their levels of BDNF would also have increased. In turn, the protein would have become a key element in the further evolution of the human brain areas important in human attributes such as complex decision making, spatial mapping, and emotional control.[36]

Running down endurance

It should come as no surprise that the Endurance Running Hypothesis is not universally accepted. The proponents of the paleo lifestyle are, as I mentioned, often critical of the carbohydrates many runners eat to fuel their exercise. Some also scoff at the hypothesis itself. Mark Sisson is the author of *The Primal Blueprint*, as well as a website called *Mark's Daily Apple*, which features a character called Grok, "an inclusive, non-gendered representative of all our beloved primal ancestors." Grok purportedly would have spent his or her (but really, from later remarks, his) days "hunting game and gathering all manner of roots, shoots, seeds and fruits for both himself and his family/small band."[37] But apparently not running. Sisson says, "We did NOT evolve to run long distances. Sure, early humans were all-around fit enough and capable of the occasional long easy jaunt after an animal, but to think that natural selection redesigned our simian shapes to run the Boston Marathon is, in my opinion, ludicrous."[38] Taleb, the proponent of fear as the motivator for exercise, says, "Nobody in the Pleistocene jogged for forty-two minutes three days a week; lifted weights every Tuesday and Friday with a bullying (but otherwise nice) personal trainer, and played tennis at eleven on Saturday mornings . . . Marathon running is a modern abomination."[39]

Among the main scholarly critics of the hypothesis are Henry Bunn from the Institute for Human Evolution at the University of the Witwatersrand in South Africa and Travis Pickering, an anthropologist at the University of Wisconsin. The two are in favor of the idea that increased ability to hunt and procure meat was instrumental in human evolution, but they are more skeptical about the importance, and feasibility, of persistence hunting by our ancestors.[40] They argue that the kind of tracking essential to successful persistence hunting is most easily accomplished in dry, open areas with little vegetation, but these were not the habitats most common where early *Homo* lived. They also doubt that those impressive tracking skills even existed in the hominins of the time, with their smaller brains and presumably more limited cognition. Furthermore, they do not think that running was necessary to scavenge carcasses, pointing out that equally effective ways of getting to and monopolizing recently dead animals might have included "power scavenging," or simply confronting a rival predator, human or otherwise, until it backed down.

Citing the foraging Hadza people of Tanzania, Bunn and Pickering say that while "they occasionally run toward perceived scavenging opportunities, the Hadza run more commonly to avoid approaching rain showers, stinging bees, and marauding elephants than to benefit any meat foraging by persistence hunting or scavenging."[41] The scientists also used the age and size of animal bones from archaeological sites, as well as the marks of tools on those bones, to deduce which individuals were selected as prey (old versus young, or large versus small) and how those prey were killed. They concluded that early humans were far more likely to have been ambush predators, like cheetahs, that lie in wait for prey and then kill it quickly, than either scavengers or persistence hunters.

Biologists Karen Steudel-Numbers and Cara Wall-Scheffler examined another assumption of the Endurance Running Hypothesis: the lack of a single optimal running speed in humans, which leads to an ability to adjust that speed for best exploitation of the prey.[42] They measured the metabolic output of nine volunteers who ran on a treadmill at various speeds, and then did further calculations about how many calories would be expended in a persistence hunt of the distance Liebenberg had found to be the average in the Kalahari hunters. The runners turned out not to be as flexible in their use of energy at different speeds as the hypothesis supposes, and the energy consumption would have been considerable, meaning that early hunters would have needed to bring down a fairly large animal to make the effort worthwhile. Steudel-Numbers and Wall-Scheffler suggest that a mixture of walking and running might have made for a more efficient persistence hunt, particularly when water was at a premium.

Bramble, Lieberman, and their colleagues counter many of these objections, pointing out, for example, that just because modern hunters use sophisticated cognitive skills and have big brains does not mean that early hunters had to have the same qualities to be effective trackers; after all, mammals like the big cats can track quite well without a humanlike brain.[43] And Lieberman's most recent study, of the heel bones in Neandertals compared to those of *Homo sapiens*, led by David Raichlen and assisted by Hunter Armstrong, lends further support to human running as an adaptation.[44] The fossil heels of ancient *Homo sapiens*, like those of modern people, had bones that allowed the Achilles tendon to become taut, while Neandertals had longer heel bones that would have been less energetically efficient during running. The efficiency of the shorter heel bones was further demonstrated by engaging volunteers to run on a treadmill while their oxygen consumption was recorded and then using an MRI scanner to measure the runners' heel bones and Achilles tendons. The individuals who ran most efficiently also had shorter heel bones, suggesting that such bones evolved in humans as running became essential.

A 2011 study by Graeme Ruxton and David Wilkinson used mathematical models to determine whether it would have been feasible for early humans to lose enough heat while running to maintain a fairly high pace during persistence hunting.[45] Taking into account the air and ground temperatures, the amount of heat lost from a mostly hairless body, and a few other variables, the scientists concluded that persistence hunting would have been effective only after humans had already evolved the ability to run long distances. This means that the Endurance Running Hypothesis might be putting the runner before his finish line, so to speak. Ruxton and Wilkinson caution that their results depend on several assumptions, as with all theoretical models, and that what we really need is more information about questions such as how dehydration affected running ability (it's doubtful that early hominins stashed water to be available during the length of a hunt), or how the relatively tall, narrow bodies of humans influence heat loss.

Were we, then, lean, tireless running machines, or robust animals that ran only far enough to find a stone for constructing a spearhead? I am convinced by the data in our bones and muscles; humans possess traits that simply work better for running than walking, and these traits aren't shared by our closest ancestors. We certainly did not evolve to be sedentary. Beyond that, it seems reasonable to suggest that we didn't run for a set period each day, like someone training for a race, but no one knows whether humans exercised like Grok. Developing elaborate scenarios about lifting boulders or walking with weights is getting into the realm of paleofantasy, and it seems like a pointless—ahem—exercise.

Greg Downey, an anthropologist at Macquarie University in Australia, suggests a sensible compromise:

> Long-range endurance running wouldn't have just been for hunting: humans cover immense ranges, even when foraging and scavenging . . . Endurance running may have been a relatively rare, but important trick, and distance running also might be useful for drop hunting (chasing animals over cliffs), for defense (run until a predator just gives up), communication, expanding our effective scavenging range, etc. . . . From our perspective from the couch, I think it's too easy to underestimate just how good we are as runners.[46]

Feet and the lawns of the Pleistocene

Regardless of whether we evolved running short distances or long ones, we certainly did not evolve to run, or walk, with shoes, and that simple observation leads to one of the biggest controversies among runners and athletes today. Part of why it is difficult for many people to imagine that the human body is "born to run" is the frequency of foot and leg pain and injuries among recreational and professional runners. A 1992 survey of the medical literature estimated that 37–56 percent of runners suffer injuries yearly,[47] and a 1989 examination of 583 "habitual runners" over the course of a year found that about half reported injuries severe enough to seek medical attention, take medication, or limit their activities.[48]

Shoe manufacturers have risen to the challenge with ever more elaborate cushioning, lacing designs, and other scaffolding intended to keep the foot stable during motion. And people are happily, or at least willingly, going along for the race. In 2009, consumers spent $2.36 billion on jogging and running footwear, up from $987 million in 1988.[49] A 2011 *New York Times* article noted that even in that year's sluggish economy, sales of running shoes were "sprinting along," having increased as much as 18 percent over the previous year.[50]

But the shoes that are selling the best are those that look nothing like traditional running shoes, with their elevated heel and thick layers of padding. Instead, the new models are minimalist, including the Vibram FiveFingers, which looks more or less like a glove for the feet, with a separate slot for each toe and a very thin sole. These shoes are designed to mimic running barefoot, which Christopher McDougall and many other proponents suggest is a more natural way to run. McDougall's website bluntly states, "The notion that the human foot is inherently flawed and automatically needs some kind of corrective device is kind of nuts."[51]

Barefoot running is a logical consequence of the idea that humans evolved to run long distances; if the whole process is artificial, then wearing special gear makes sense, just as it does for ballerinas donning pointe shoes; they couldn't perform as required without assistance. But if running is how we are built, then the fancy equipment may be superfluous, like attaching helium balloons to a bird's wing.

McDougall points to the Tarahumara of Mexico, who run either barefoot or with very thin sandals for up to 100 miles over rocky terrain. He and others also note that little evidence exists that wearing shoes actually prevents injury.[52] One study published found that shoes costing more than $90 were associated with more than two times as many injuries as the cheapest models.

As McDougall and others, including Lieberman, who is also a barefoot-running aficionado, are quick to point out, the real difference is not so much the shoe as the technique one uses while wearing (or not wearing) it. Shod runners tend to strike the ground more heavily, with their heel first, while barefoot runners land more springily, on the mid- or forefoot. It is this gentler method that seems to be the key to avoiding injury. In keeping with this idea, a 2011 study of women runners who tried out minimalist shoes found that if they didn't change their form to land on the ball of the foot, they still had hard landings and high impact.[53]

Lieberman and a host of colleagues set out to examine this idea by comparing the gaits of both barefoot and shod runners in Kenya and the United States.[54] They calculated the collision strike force for five groups of people who all ran at least 20 kilometers each week: athletes from Kenya's Rift Valley, who had grown up without shoes but now ran with them; two groups of schoolchildren from the same place, one of which had never worn shoes and one of which were mostly shod; American athletes who had once run in standard shoes but who now ran either with minimalist shoes or barefoot; and American athletes who routinely wore shoes. The barefoot runners tended to hit the ground with the forefoot, with occasional midfoot or heel strikes, whereas the shod runners were primarily heel strikers, with more impact and jarring of the body. It's that forceful collision, Lieberman and his coworkers claim, that causes runners the most harm. If a runner takes off his or her shoes but doesn't adjust technique, injuries are still likely.

Interestingly, the forefoot- and midfoot-striking runners in Lieberman's research had softer landings even on hard surfaces. In many discussions of barefoot versus shod running, contributors on either side point out that going without shoes was fine back when our ancestors were doing it, but we did not evolve to run on concrete or other hard surfaces. But as McDougall states on his website, "What soft, grassy fantasy land did they come from? Check out the sun-baked African savannah sometime: hard as cement. Or the stone trails of Mexico's Copper Canyons, or the packed dirt roads of Ancient Greece."[55] Certainly some runners simply want a bit of protection to keep their feet away from broken glass or burning cigarette butts, unarguably not Pleistocene hazards. Nevertheless, the argument that barefoot running was all well and good back in the day when the soft Paleolithic soil yielded under our feet, but not now, is, well, a paleofantasy.

A 2011 review from the *Journal of the American Podiatric Medical Association* was fairly noncommittal on the topic of barefoot running but noted that "many of the claimed disadvantages to barefoot running are not supported by the literature."[56] The authors found no increase in shin splints and other injuries among barefoot runners, for example.

People who take up barefoot running seem to become evangelical about it, touting the joy of feeling the trail under their bare feet. One commentator on the *New York Times Well* health blog said, "Running on a regular basis may not work out for you, but never forsake the barefoot experience . . . except in restaurants and the Opera/ballet, and presidential luncheons."[57] In the meantime, sales of minimalist "barefoot style" shoes were up 283 percent in 2011 over the previous year, and the arguments continue to rage. As another *Well* blog reader said, "I'd like to see scientists measure levels of smugness and self-righteousness in barefoot vs. shod runners."[58]

The real problem with trying to characterize the natural foot or its gait, however, is that feet, like other components of our bodies, do not develop in a vacuum. If you put those adorable tiny shoes on a toddler's feet, the shoes alter the growth pattern of those feet. If you don't put shoes on them, the growth pattern is still altered—by the surface the child walks on, by the amount of time the child spends doing different activities, and so on. Downey points out that feet can perform all kinds of tasks, including painting or typing on a keypad, by people whose arms or hands are disabled. "The fact that skills like foot painting or feeding oneself with one's feet are rare does not mean that our feet were not 'designed' to do them," he says.[59] Even with the same genes, one person's foot can become adept at running, another's can learn to excel at playing an instrument, and yet another's can become atrophied if it is bound according to ancient Chinese practice. Our genes had this responsiveness in the Pleistocene as well, making it impossible to isolate a single "most natural" exercise pattern.

Genes, muscles, and a race for the future

Finally, what about our running future? Is our ability to exercise itself able to evolve? Scientists are finding intriguing hints of the kind of genetic variation that is the fodder for evolutionary change. Assuming, of course, that we can overcome the sedentary siren call of our couches.

Athletic ability hasn't shown the kind of clear-cut, widespread genetic shifts in distinct populations that lactose tolerance has, but people obviously differ in their ability to perform physical tasks, and at least some of that ability is founded in the genes. And while parents who gaze admiringly on a tot as she flings strained peas have always dreamed of the WNBA, now they can actually test the genes of their tiny would-be athlete to see if those dreams are likely to be realized. Or, to be more accurate, a test is available. Whether it really guarantees a career on the court is another matter.

The gene in question is called *ACTN3*, and it is responsible for the production of a protein that controls fast-twitch muscle fibers. Muscle tissue exists as strings of molecules that can contract quickly or slowly; the fast-twitch type is important in activities requiring bursts of power, such as sprinting. Some people have a variant of the gene that disables such fast-twitch fibers, and if they have two copies of the variant, they tend to have slightly weaker muscles and run slightly slower, though the variant is also more likely to be found in endurance athletes such as marathon runners. The gene variant influences the muscles by making them more efficient at aerobic metabolism, the kind of metabolism used during endurance activities. The variant is present at different frequencies in people from different parts of the world, with, for example, about 18 percent of those from European ancestry but only 10 percent of Africans possessing it.[60]

Australian researcher Daniel MacArthur and his colleagues studied *ACTN3* both in humans and in mice that they genetically engineered to lack the protein.[61] The mice without the protein could run 33 percent farther than mice with the more normal gene configuration, which presumably would correspond to mice that were better at running the rodent equivalent of a marathon. Mice with the variant also had 6–7 percent less grip strength, though they were still well within normal mouse capacity.

The scientists then reasoned that if variation in the type of *ACTN3* present influenced running performance, and the percentage of people with the variant differed across the globe, it should be possible to determine whether *ACTN3* evolved recently because of natural selection on muscle ability. To test their hypothesis, they did the same analysis of the genes surrounding the *ACTN3* gene that I described for lactose tolerance in Chapter 4, looking to see whether the neighboring genes were swept along as selection for the chunk of DNA containing *ACTN3* was favored. Indeed, the gene variant seems to have become much more common over a relatively short period in European and Asian populations.[62]

As might be expected, Olympic sprinters were less likely to have the variant than Olympic endurance athletes, though the athletes did not show a single genetic profile even for *ACTN3*. Studies of a handful of other genes have suggested a disproportionate frequency in elite athletes, leading a group of Spanish researchers to ask whether there is an optimal genetic profile for such people. Led by Jonatan Ruiz and Alejandro Lucia, the Spanish team examined 46 "world-class endurance athletes," including both endurance runners and professional cyclists, all of whom had participated in either an Olympic final event or a Tour de France, and 123 "non-athletic (sedentary) controls."[63] Although the athletes tended to be more likely to have favorable genes for performance, no one had the perfect profile, and the authors suggest that genetic variants "yet undiscovered as well as several factors independent of genetic endowment may explain why some individuals reach the upper end of the endurance performance continuum."

This conclusion should come as no surprise to anyone who has trained for an athletic event or watched someone else do so. Which brings us back to the idea of whether it is worth testing little Sally to see if she should be sent to Olympic summer camp. In MacArthur's opinion, the answer is a flat-out no. In an article on Wired.com, he says, "**This is not a test that can tell parents whether or not their kid will be able to become an elite athlete in general.** No matter what your *ACTN3* genotype is, there is a wide range of sports that you could theoretically excel in" (emphasis in the original).[64] Too many genes, not to mention environmental factors, contribute to athleticism to make testing for a tiny variation productive.

What about the runners, Olympic-caliber or not, of tomorrow? As Downey argues, and I agree, it is hard "to talk about . . . what the human body is 'designed' to do."[65] French Nobel Prize–winning geneticist François Jacob famously said, "Nature is a tinkerer, not an engineer."[66] He meant that evolution can work only with the parts at hand, rather than inventing the best possible solution to a problem. Hence, people in the Pleistocene may well have had knee trouble, because our knees are jury-rigged joints that have been modified from our quadruped ancestors. That doesn't make bending down unnatural. And we use our limbs, and our bodies, for a wide variety of activities, depending on our place in history and on the planet. It is therefore futile to look for the single best type of exercise, given our evolutionary heritage, though it's a safe bet that we would be healthier if we got up off the couch.

7

Paleofantasy Love

What is our true sexual nature? Humans have highly demanding, precociously born babies; although a mother chimp or gorilla can raise an infant on her own, in all human societies women usually have help with their offspring, often in the form of the father. One could take this to mean that monogamy is both natural and necessary. But a great many popular and scientific sources disagree. Some suggest that men and women are in conflict, with naturally faithful women fighting a losing battle to keep men from straying. As one commenter on an article titled "Is Infidelity Natural? Ask the Apes"[1] says:

> Men are wired to spread seed far and wide. All men know this, we get hormones that run through our bodies that do this. There is no denying that at all, and it would be stupid to deny it. Men are also logical and can hold those feeling [*sic*] back. It doesn't always happen.[2]

Alternatively, the best-selling book *Sex at Dawn* would have it that both men and women are sexual gourmands, with multiple

(© Kim Warp/The New Yorker Collection/www.cartoonbank.com)

partners the real norm and monogamy a miserably failed experiment. According to authors Christopher Ryan and Cacilda Jethá, "The campaign to obscure the true nature of our species' sexuality leaves half our marriages collapsing under an unstoppable tide of swirling sexual frustration, libido-killing boredom, impulsive betrayal, dysfunction, confusion, and shame."[3]

On Cavemanforum.com, readers chime in:

> It's actually also true than [*sic*] girls are more likely to settle for the less alpha-type guys, but if they are going to cheat they do it with the more macho guys. We're biologically wired that way. Blame evolution.[4]

> The paleo norm was for less than half the men to reproduce. Having 80–90% of men reproduce is a very recent phenomenon (even closer in time than agriculture). Lifetime Monogamy is as artificial as bread or corn syrup.[5]

What all of these views have in common is an evolutionary perspective. They seek to explain our modern behavior in terms of our past, recognizing that how we mate and have babies is at the heart of who we are. Sex is the way in which we can seem most like our animal ancestors and relatives; all creatures reproduce in some fashion, and we obviously are descended from primates that chose mates, gave birth, and raised their children. What is not so clear is how they did it, and by extension what behaviors and roles we have inherited from our ancestors.

Perhaps not surprisingly, many of the paleo-lifestyle advocates point wistfully to a time when "sex was a lot more egalitarian and promiscuous than you'd think it was."[6] One woman confessed that, since eating a paleo diet, "I'm VERY attracted to more aggressive (but still respectful), more capable men, strong of intellect and body, physically larger, hair on their chests, who would make good protectors and providers—not so much financially, but in the home."[7] Exactly how a change in diet leads to a preference for more hirsute partners is not clear.

We may not really believe that men are from Mars and women from Venus, but we have a lot of opinions about what each sex can do or is likely to enjoy, from reading maps to watching tearjerker movies. Men and women are often supposed to differ even more in their sexuality, whether that means the kind of partner they want, their sexual appetites, or what they think a good marriage looks like. Evolution has been invoked to explain all of these characteristics—maybe not the movies per se, but supposed sex differences in traits, ranging from the capacity to appreciate nuances in relationships or have empathy for the travails of others, to the ability to perform spatial reasoning tasks, to a love of shopping for shoes, have been attributed to our history as hunter-gatherers.

As with the question regarding other aspects of human behavior, however, how much do we know about the sex lives of our ancestors? And how much of that behavior is still manifested today, in a world with speed dating and sperm banks? Is it instructive to look at our ancestry, or will we simply see what we want to see?

Darwin, peacocks, and pipefish

To understand the evolutionary basis for sex differences, we need to pay a brief visit to Charles Darwin. Although he is best known for his theory of natural selection as it applies to the origin and diversity of species on Earth, Darwin was also extremely interested in sex. One might almost say he was troubled by it, not necessarily in his own life (though some biographers have speculated about problems in his relationship with his wife, Emma, as well as his Victorian-era prudery), but because many things about sex in animals seem perplexing.

Take, for example, the peacock, with his splendid shimmering train of green, blue, and gold. Anyone who has seen a peacock displaying at a zoo will have no difficulty imagining that the huge fan of feathers impedes the male's movements, and would make it difficult for him to escape, say, a tiger creeping up in the forests of the peacock's native India. The females lack this encumbrance, as is the case for many kinds of animals: though exceptions occur, males are the ones with the elaborate plumage, colorful patterns, and loud songs—all characteristics that require a fair amount of energy to produce and that render their bearer more conspicuous to predators. In short, these traits seem disadvantageous, and hence would not be expected to evolve under natural selection.

Darwin's solution was that another process besides natural selection was at work, called sexual selection. Just like natural selection, sexual selection causes some organisms to leave more genetic representations of themselves than others do, but instead of operating on characteristics such as a greater ability to blend in with the bark of a tree and hence avoid detection by a hungry bird, sexual selection is all about who gets more and/or better mates. If a female peafowl (the correct term for the species; "peacock," strictly speaking, refers only to the male, while "peahen" is the word for the female) is more likely to mate with a male bearing an ornamented train than one with more modest plumes, the genes conferring the lovely feathers are more likely to appear in succeeding generations. The process works even if the traits hamper survival, though if the advantage in mating is outweighed by a disadvantage in survival, sexual selection won't continue to act to exaggerate the trait.

Sexual selection is thought to account for many of the differences that we see between males and females: male deer with enormous racks of antlers, brightly colored birds of paradise, and frogs that croak the night away. It also explains why it is mainly the males that sport fancy advertisements, and not the females. Darwin's original idea relied on the aforementioned Victorian notions of coy females needing to be persuaded by macho males, and it has been modified by his successors.

In my opinion, the best framework for understanding sexual selection comes from the more modern interpretation by Robert Trivers, an evolutionary biologist at Rutgers University. In a book chapter published in 1972, Trivers pointed out that although both sexes devote a great deal of effort, time, and energy to perpetuating their genes, the way in which they do so often differs.[8] Evolution means changing the genes that occur in the population, which means that individuals who have higher reproductive output are more likely to be represented. What limits that output? For females, it's the number of offspring they can produce and, in some cases, rear, because by definition the female is the egg-producing sex and the eggs, once fertilized, are usually (though not always) cared for by the mother. Because eggs, and embryos and fetuses, are expensive and time-consuming to manufacture, even if females operate at maximum capacity their production is fairly small.

Imagine, for example, the highest number of puppies or kittens a female dog or cat could have in her lifetime, assuming that she mated every time she came into heat. That number, while obviously higher than the number of children a human female could bear over her lifetime, is still vastly smaller than the number of young that, at least in theory, could be sired by a given male of the species, since his investment, as Trivers would refer to it, is only the relatively smaller amount of time and energy required to mate with the females. Given the opportunity, a single male could inseminate many if not all the females in a population. Therefore, a male can win much bigger, evolutionarily speaking, than a female, because her upper limit is smaller. But of course, one male's win is another's loss, and in many species, most males do not mate with even a single female. Sexual selection therefore favors males that are able to compete with members of their own sex to get the opportunity to mate at all, because the stakes are high.

Females, on the other hand, will often do better from an evolutionary perspective if they mate with only the highest-quality partner, because a male that either has genes conferring a greater ability to survive or provides for the offspring directly—say, by feeding them—will make it more likely that her own genes persist in future generations. With this dichotomy in mind, many evolutionary biologists talk about competitive males and choosy females. Exceptions abound, of course, and Trivers himself emphasized that it's the level of investment that is important, not the sex of the individual doing the investing. If, as is the case in some insects, the male must present the female with a nutritious package that he produces with his own body fluids before she will accept him as a mate, his investment becomes quite high, and males of species that offer these "nuptial gifts," as they are called, are quite choosy about their mates. Seahorses and their relatives the pipefish are also good examples of this reversal of the usual state of affairs, with males receiving the eggs of females in their pouches so that they become pregnant and give birth to tiny replicas of themselves after the young develop.

Some researchers and popular writers took the idea of selection favoring different behaviors in the sexes to mean that women are expected to be rather unenthusiastic about sex itself, or at least to want it only as a means to gain a male's investment in children, while men are expected to enjoy sex for its own sake. Presumably, the thinking goes that if men invest only sperm in mating (the more the merrier, as it were) while women have to weigh the consequences of their pleasure in a potential pregnancy, selection should favor men who seek out one encounter after another. Ryan and Jethá, the authors of *Sex at Dawn*, are among the most explicit about this claim, and the most eager to debunk it. They refer to Darwin's "anti-erotic bias" and, more directly, claim that "Darwin says your mother's a whore."[9]

Evolutionary psychologists, who extend some of the principles of evolutionary biology to the mental and behavioral adaptations of our own species, also come in for criticism by Ryan and Jethá, who note that many works coming from that field also assume that the female libido is sluggish, at least in comparison to that of the male. My own opinion is that while it's certainly true that biologists, like everyone else, have brought their own biases to the study of human sexuality, whether in early evolution or not, the differential investments of the sexes into offspring do not automatically lead to a reduction in female lust. Just because the consequences of sex are different for men and women does not mean that either one enjoys it less.

Who, really, is your daddy?

Another reason males are said to be more likely to roam is central to the difference in how males and females reproduce. With rare exceptions, any offspring produced by a female is guaranteed to carry half of her genes, which makes any investment into that offspring a sound one in evolutionary terms. The same is not necessarily true for a male; he cannot be sure, anthropomorphically speaking, that a given offspring is his, because the mating happened days, weeks, or months before the young are hatched or born, and the female could have mated with another male in the meantime.

This disparity between the sexes means that what is called confidence of paternity can be rather low in some species, especially those in which females mate more than once before their eggs are fertilized. If a male is not the genetic father of the offspring a female produces, he is better off, evolutionarily speaking, finding another female to mate with than sticking around to help rear the young of his first mate. Females, on the other hand, would benefit from having a male's help in taking care of their babies, regardless of whether he is their genetic father. Hence, the story goes, females are more inclined to monogamy and males to playing the field. And indeed, in many species the males perform a variety of activities that seem to have evolved to increase the likelihood that they are indeed the only one to have mated with a female; in bluebirds, for example, males stick close to the female during the critical few days surrounding ovulation, chasing any other suitors away.

We humans seem to be no less concerned than the bluebirds. So-called paternity fraud, with women fooling men about their relationship to the women's children, has become a staple of talk shows and TV crime series. Billboards provocatively ask, "Paternity questions?" and advise that the answers are for sale at your local pharmacy in the form of at-home DNA paternity tests. Some fathers' rights groups in Australia have called for mandatory paternity testing of all children at birth, with or without the mother's consent or even her knowledge.

And people seem quite convinced that all this vigilance is necessary. When asked to estimate the frequency of such misassigned paternity in the general population, most people hazard a guess of 10, 20, or even 30 percent. The last number came from a class of biology undergraduates in a South Carolina university that I polled a few years back. I pointed out that this would mean that nearly twenty people in the class of sixty-some students had lived their lives calling the wrong man Dad, at least biologically. They just nodded knowingly, undaunted. Even scientists will often respond with the 10 percent figure, as a geneticist colleague of mine who studies the male sex chromosome—and knew the real answer—found when he queried fellow biologists at conferences.

The truth, however—insofar as we can tell—is much less sensational. The most unbiased research suggests that the real incidence of misassigned paternity in Western countries hovers around 1 percent, with a few studies pushing that number to 3 percent or nearly 4 percent. Even at the high end, that's only one-tenth as common as conventional wisdom would have it. Obtaining a truly unbiased estimate is difficult because most people undergo paternity testing only if they have a reason to suspect a discrepancy between the purported father and the genetic one. As a result, using data from the companies that sell the at-home tests, for example, is certain to yield an overestimate of misassigned paternity.

A handful of studies that get around this problem do exist, mainly in the medical literature. Most of these studies gathered information on the parents of children with genetic disorders like Tay-Sachs disease or cystic fibrosis, in which the child must inherit a copy of the defective gene from both parents to show the disease. When large numbers of families are surveyed for such research, a certain proportion of fathers turn out not to have the gene that their purported child inherited, thus yielding the figures of 1–3.7 percent. Higher numbers, particularly the often-cited 10 percent, apparently come from more biased populations, or, more likely, simply turn out to be urban legend, akin to cell phones being able to pop popcorn.

Interestingly, no one seems to question the apparent contradiction between the stereotype of faithful females and all this suspicion. The problem is that if men are going to have urges to stray, they have to stray to, or with, a partner—it takes two to cheat, of course. As Hester and her scarlet letter would attest, however, we have a double standard about philandering. Regardless of who is to blame, why we are so ready to believe an inflated figure of our own infidelity? Perhaps our cynicism feeds into already-held beliefs about the nature of male and female sexuality, as evidenced in the online comments quoted earlier in this chapter. Of course, we really don't know how low, or high, confidence of paternity might have been in our ancestors, but these modern results suggest that men might not be such philanderers, or women such deceivers, as the more pessimistic among us would have us believe.

Not-so-modern love

How does the extraordinary fatherly capacity of seahorses or the burden of the elaborately ornamented peacock apply to humans? People looking to understand what our ancestral mating behavior was like use the same sources as those examining other aspects of human evolution: other primates, especially chimpanzees and bonobos; modern hunter-gatherer societies, along with written records of marriage patterns in ancient civilizations; and the evidence contained in our own bodies, which bear witness to the selection pressures that acted on men and women in our evolutionary history. All of these reference points have their shortcomings, but each is also valuable, so it is worth examining them in turn.

First our primate relations. The great apes—gorillas, chimpanzees, bonobos, and orangutans—vary quite a bit in their mating patterns. Gorillas live in groups with a single dominant male (the silverback, so named because age has grayed his fur) and multiple females, while orangutans are solitary for most of their lives and both chimps and bonobos have complex societies with many individuals of both sexes. Chimpanzee society is marked by pronounced male aggression, even violence, while bonobos are more likely to share food with other group members. Males and females in both species will mate with multiple partners, and none of our closest relatives show long-term pair bonds.

Because the chimps and bonobos are the animals with whom we most recently shared a common ancestor, people have long been interested in whether some of our sexual proclivities, such as fidelity or the lack of it, or the preference for a particular kind of sexual partner, might have its origins in their behavior. Whether finding such common ground means that the aforementioned proclivities are "natural" (whatever that means), much less genetically determined and unable to be altered, is a separate question. For now, let's just see how our sexuality could be mirrored in that of our kin.

Virtually everyone who has studied bonobos is struck by the primary role that sexual behavior plays in their lives; they tend to settle conflicts with sexual activity, both between males and females and between members of the same sex. Like the chimpanzees, bonobos live in large and somewhat fluid groups containing males and females, but unlike chimpanzees, bonobos tend to be less violent in their day-to-day behavior, and female bonobos can dominate males and chase them away from food at least some of the time.

Scientists have used both chimps and bonobos as models of what early hominin sexuality might have been like for many years, although the latter have been well studied since only the late 1970s, while researchers began documenting chimpanzee social behavior several decades earlier. The emphasis on aggression in the chimps and open sexuality in the bonobos has not gone unremarked, with the frequent claim that "chimpanzees are from Mars, bonobos are from Venus." Eminent primatologist and author Frans de Waal has written extensively on how we can—and can't—use the bonobo sex life to understand our own, pointing out that while they "provide a concrete alternative to 'macho' evolutionary models derived from . . . chimpanzees," the bonobo's society is very different from our own, and we do not know how closely it resembles that of our early ancestors.[10]

De Waal goes on to speculate, "Had bonobos been known earlier, reconstructions of human evolution might have emphasized sexual relations, equality between males and females, and the origin of the family, instead of war, hunting, tool technology, and other masculine fortes."[11] This is one possibility; the other is that if we had known about bonobos earlier, we would have characterized them as more violent and warlike than we do now, simply because anthropologists and primatologists in the 1960s and '70s were disposed to emphasize male aggression, which bonobos do exhibit, albeit to a lesser extent than chimps do.

In their book, Ryan and Jethá are very pro-bonobo, appreciating the face-to-face kissing that the species exhibits, along with the slow development of its infants, like that of humans but different from chimpanzee development.[12] They also note that both people and bonobos have sex under circumstances other than procreation, such as when resolving conflict or cementing bonds between individuals, and they cite this similarity as grounds to believe that the bonobos are a more accurate representation of our ancestral state than are those no-nonsense chimps. They then go on to promulgate a rather orgiastic view of human sexuality, saying that our monogamous woes arise from an uphill, and ultimately doomed, battle with our more bonobo-like desires for multiple sexual partners. An abundance of casual sex, or as Ryan and Jethá put it, "Socio-Erotic Exchanges," could then function to make the wheels of society run more smoothly, though I have trouble imagining the kind of genital-genital rubbing that is commonplace among female bonobos ever catching on at book clubs.

I will return to the idea of monogamy and its role in our evolution later in the chapter, but for now I want to comment on whether the bonobos, or any other living primate species, are a realistic model for our earlier behavior. It's true that the bonobos and the chimpanzees are our closest relatives on Earth, but at the same time, we have not shared a common ancestor for at least 5 million years, so more than enough time has elapsed for selection to act separately on each of the three species. The bonobos may share nonreproductive sexual activity with humans, but vervet monkeys, a species much less closely related to us than are any of the apes, also have sex outside the time when a female can conceive. There is no *a priori* reason to expect that any particular aspect of our sexuality has been preserved from one ancestor and not another.

Why do the gender relations of our primate relatives matter at all? As anthropologist Craig Stanford points out, "The behaviors at the heart of the chimpanzee-bonobo interspecific variation—sexuality, power and dominance, aggression—are those that also lie at the center of the debate about human gender issues and what molds our own behavior."[13] He also extended feminist anthropologist Sherri Ortner's contention that "men are to women as culture is to nature"[14] by wondering if "chimpanzees are to bonobos as men are to women."[15] In other words, even now that we know about bonobos, males and females are still seen as different, and the genders are still stereotyped, but we get to pick whether we like the old male version with the war toys or the new female one with lesbian sex and food sharing. Either way, however, we are simply imposing our preexisting biases ("free love is natural," "males are violent brutes") on species that are complex in their own way, not as caricatures of people. A better idea is to figure out what the animals are like without using them as role models.

What's more, new information on the role of evolutionary history in primate social life suggests that an important gap lies between us and our ape relatives. A 2011 study by Susanne Shultz, Christopher Opie, and Quentin D. Atkinson traced the evolutionary history of social behavior and found that genetic relatedness was more important than environmental conditions in determining whether a given species lived in pairs, small groups, or large groups.[16]

History is important because if one is interested in understanding the evolution of any trait, whether social-group composition or a tendency to eat leaves, and that trait occurs in a number of species, the first question is whether each species inherited that trait from the same common ancestor, or whether it arose independently multiple times because of the same selection pressures. In the case of primate social-group structure, the conventional wisdom had it that our more distant ancestors lived in simpler societies with just a pair of animals or perhaps a family group. Then, larger, more complicated groupings evolved. The details of group structure—whether solitary or in single- or multi-male societies, for example—were thought to depend on the location of food in the environment and other ecological variables.

Shultz and colleagues, however, were able to place published information about the social organization of 217 living primate species in the context of the genetic relationships, and hence the evolutionary history, of those species. They discovered that social organization tended to be more similar among closely related species than would be expected by chance, which means that genes, not ecology, play a big role in the kind of society a species exhibits. Perhaps even more interesting, they found that, about 52 million years ago, primates went from a solitary life to fairly unstructured groups, and then to more stable ones. Then, both pair-bonded societies and single-male harems, like those seen in gorillas, emerged roughly 16 million years ago, rather than having a more linear evolution with groups always becoming more complex.

The age of social systems is significant in the context of the evolution of our sex lives because we know that the last common ancestor of humans and chimpanzees lived about 5–7 million years ago. Since modern chimpanzees and bonobos live in stable communities with multiple males and females, pair bonds—the early version of monogamy—must have emerged sometime after we all diverged from this more recent ancestor. When that happened is still a mystery, but the new data suggest that human mating patterns evolved on their own for some considerable time after that divergence.

Genetic and linguistic signatures of mating history

The second source of information about our sexual natures is the life of humans in contemporary cultures. Human beings are unusual among animals because of our frequent, though by no means universal, monogamy; a number of species pair up for the duration of a breeding season, and a small handful mate for life, but by and large, animals tend to display a variant on the multiple sexual partner theme. And indeed, sexual selection theory suggests that because males so frequently gain by attempting to mate with multiple females, we expect "polygyny" (the scientific term for a single male mating with more than one female) to be the most common mating system. A few species do show the counterpart to this system—polyandry—in which one female mates with several males at the same time, but it is limited to situations in which a female can mate and produce offspring with one male, leave those young with him, and then proceed to the next. Polyandry has been observed in human societies as well, but again only under very limited circumstances.

Where, then, did our monogamy come from? Is it a cultural artifice that denies our basic nature, as Ryan and Jethá would have it, dooming countless men and women to lives of guilt and secret philandering? Or is it an adaptive part of human society? In a volume of multicultural perspectives on the family, Bron Ingoldsby claims, "Monogamy is certainly the most common of the marital types; the generally equal sex ratio throughout the world sees to that."[17] But this is specious reasoning; after all, males and females occur in roughly equal numbers in many animal species, but they exhibit a variety of mating systems, with many individuals, usually males, simply not mating at all. The real question is, what happened in human evolution? Or put another way, from a blog entry by historian of science Eric Michael Johnson, "Were our ancestors polygamists, monogamists, or happy sluts?"[18] People have argued the question from all sides, but we are only now gathering the data that allow us to test it.

Polygyny is still found in many historical and modern human groups, with wealthier men able to have multiple wives, while poorer ones can afford to have only one or stay single. Monogamous marriage as we now know it was thought to have become more prevalent after societies became more complex and agricultural and less nomadic. Laura Fortunato, an anthropologist at University College London, analyzed marriage patterns from Eurasia, using a tree of cultures that mimicked an actual evolutionary tree such as that showing, say, the relationships among various members of the cat family, with lions and leopards more closely related to each other than either one is to lynx or pumas. In Fortunato's case, she incorporated the group's historical relationships based on language, so that, for example, Portuguese- and Spanish-speaking peoples were seen as more closely related to each other than either was to Lithuanians.[19]

The results of the tree construction were surprising. Of the twenty-seven societies sampled, eighteen, or two-thirds, were monogamous, while the remaining one-third were polygynous. This finding in itself was not unexpected, but more interesting, monogamy appeared to have arisen well before modern record keeping. In other words, at least some of the first settlers emerging after our hunter-gatherer nomadic ancestors appear to have been monogamous. This puts to rest the argument that monogamy is a recent invention that required the development of a more industrialized way of life.

A different line of evidence, however, suggests that the pattern of multiple partners, or at least multiple females mating with a single male, was also a part of our history, just as Ryan and Jethá would have it, albeit with a twist. New information about the human genome allows scientists to use our genes to look into the past. Specifically, to estimate ancient mating patterns, researchers can study the way genes vary on the sex chromosomes and then compare that variation to the way genes vary on other chromosomes.

Women have two X chromosomes, while men have only one, which means that mothers always pass an X on to their children but men pass one on to only their daughters. This disparity means that women contribute disproportionately to the genetic diversity on the X chromosome, compared to men. And what *that* means is that if a society is polygynous, so that a relatively small number of men have multiple wives, those few men will reduce the overall genetic diversity in their offspring—except for the X chromosome. The men contribute an X to their daughters, but the women also contribute an X to the daughters, as well as one to the sons. Thus, although the fathers reduce the diversity overall, the mothers should maintain the X chromosome diversity. Therefore, a comparison of the genetic variability on the X chromosome compared to the rest of the chromosomes should reveal higher genetic diversity on the former—which is exactly what Michael Hammer and his colleagues found in samples from populations around the world. They concluded that polygyny was a part of our history, and the evidence is written in our genes.[20]

What about that twist? The idea that more women than men left genes in succeeding generations is supported by these new data. But genetic variation in a population is influenced by more than just how many males mated with the available females. Another important contributor is how much each sex moves from its birthplace: new immigrants contribute new genes, while those who stay home and marry neighbors or especially relatives (even distant ones) will reduce the overall genetic variability. In some societies, men stay and women go to the areas where their husbands were living; in others the reverse occurs.

Hammer's study and others similar to it could track the genetic signature of polygyny back to the start of agriculture, about 10,000 years ago. Anthropologists believe that around that time, along with eating grain, people began to have more patrilocal residence patterns, with women leaving the places where they were born. Such societies are more likely to be polygynous, because multiple women can all go to the same man's residence, and they would also reinforce the pattern of X chromosome variation that Hammer and his coworkers found. But such an increase in polygyny would have been a fairly recent phenomenon, rather than one established deep in our history, which means that monogamy could have been ancestral to humans. Other studies have found results that differ from Hammer's, with the discrepancy now being attributed to differences in the timescale at which the data are analyzed.

The paleofantasy of a caveman past in which a few dominant males held sway and women meekly served them is as unrealistic as one in which we all paired up and never strayed. Humans have successfully reproduced under a variety of mating systems, depending on where on the planet and when in our history one looks. As with diet, as with exercise, as with all the other features of our biology that people want to make into a single "natural" way—we don't have just one natural pattern of the sexes.

Hunting, gathering, and sex

Whether our ancestors were monogamous or polygamous, another hallmark of humanity is a sexual division of labor, with men and women often performing different tasks in a society. That division is frequently linked to other differences between the sexes, like the purported lack of mathematical ability in women, and to the evolution of the nuclear family and (once again) the knotty issue of infidelity. So, do we really come from a 1950s-style family structure, where men went out and brought home the mastodon meat while women raised children and dug for roots in the soil? Or—and you can probably guess what's coming here—is the real story more complicated?

Thus far in this chapter I have been ignoring an important, perhaps the most important, component of the story about sex and evolution: the aftermath of the act—namely, the children. Among animals, monogamy usually evolves when the offspring require so much care that one parent cannot rear them alone, and humans fulfill this requirement in spades. As the noted anthropologist Sarah Blaffer Hrdy painstakingly recounts in her books *Mother Nature* and *Mothers and Others*, and as I will discuss in Chapter 8, human babies are almost breathtakingly demanding. They need food provided to them well past the time of weaning, and they need to be protected from the elements and predator attacks. From a cold-blooded, practical perspective, it is many years before they can start returning on the investment made in their care by contributing resources to the family.

How do these time- and energy-sucking little creatures manage to survive? Many biologists have reasoned that the answer lies in family structure, and particularly in the fathers of the children. Starting with Darwin, scientists have theorized that prehistoric men hunted and brought the food home to their mates and offspring. In return, each woman stayed by the fire and remained faithful to her man, guaranteeing him confidence of paternity. The evolution of our big brains was also part of this scenario, because selection would have favored smarter men to be better hunters, who were then more likely to successfully provide meat for their families. According to this hypothesis, sexual division of labor is thus essential to our human uniqueness because it drove the evolution of intelligence, creating a feedback loop in which ever-smarter individuals reinforced the social system.

Hrdy calls this exchange of meat for fidelity the "sex contract," and versions of it have remained a part of the story of our evolution for the last several decades. Hrdy's interest in it pertains mainly to how it affects child-rearing and the human family, which I will discuss in more depth in Chapter 8, but the sex contract is also relevant for understanding gender differences and our sexual nature itself.

For many reasons, hunting is seen as a lot more glamorous, and hence important, an occupation than gathering, and for much of the twentieth century, bringing home the meat was viewed as more central to human evolution than picking the berries. Putting hunting, and males, front and center in ideas about the evolution of gender roles also meant that women's work was not seen as particularly valuable to the group. Anthropologist Lori Hager suggests that the early models of Man the Hunter were popular in part because they validated the way Western families were structured in the 1940s–60s.[21] Even today, most museum dioramas and other illustrations of prehistoric family life depict a man setting out with a spear, or holding a captured rabbit, while a woman sits by the campfire, an infant at her bosom. The implication is that women stay home and care for the children, while men go out and bring home the bacon, or mammoth meat.

Starting in the 1970s, a number of mainly women anthropologists, most notably Adrienne Zihlman from UC Santa Cruz, began to question this macho perspective. Zihlman and several others discovered that in contemporary foraging peoples, women's gathering often provided the bulk of the nutrients consumed by the group, and furthermore, that in some cultures, such as some Australian aboriginal tribes, women hunted as well. But the "Man the Hunter" model has persisted; Zihlman suggests, "It came to stand for a way of life that placed males center-stage, gave an evolutionary basis for aggressive male behavior and justified gun use, political aggression, and a circumscribed relationship between women and men as a 'natural' outcome of human evolutionary history."[22]

Zihlman and others since the 1970s have promoted a different version of human evolution, dubbed "Woman the Gatherer," focusing on female contributions and women's lives beyond child rearing. Other ideas have been proposed over the last few decades; for example, food sharing, particularly among non-kin, is sometimes viewed as an important component of early human evolution, because it sets the stage for complicated social exchanges among groups beyond the family, as I will explain. In addition, the relative roles of hunting and gathering in our ancestors are still being debated by anthropologists.

Rebecca Bliege Bird of Stanford University noted that among modern foraging peoples, although the food items acquired by men and women differ, women tend to bring back abundant, small, and low-risk foods, like shellfish or berries, while men obtain the rarer and harder-to-catch items, like sea turtles or deer.[23] Is this division of labor simply a matter of everyone doing what he or she does best to support the household? Maybe not; by concentrating on the more risky items, the men may simply fail to hold up their end of the bargain. Even when men procure plant foods, as with the yams that Melanesians use as a staple, according to Bird, men "compete to grow outsized roots, sometimes spending days preparing a single hole, while women plant dozens of 'table' yams in smaller holes."[24] The resulting men's contributions, though impressive, are apparently used in the Pacific equivalent of county fairs, and end up being divided to use for future planting rather than eaten.

Bird examined the possibility that women must trade off the need to take care of their children with the demands of foraging, since hunting big game is not compatible with caring for infants and toddlers, but while the need for work-life balance seems as real for the Aché of South America as it is for Manhattanites, that trade-off is not the whole answer. Women who are past reproductive age do not seem to spend more time hunting than do younger women, at least in some foraging cultures, so there must be more to the story than the demands of child care.

Furthermore, humans from many societies worldwide share food, not only with their children and mates, but with other members of society. Such generosity can play an important role in oiling the wheels of social interaction, but it needs to be balanced with what each sex is capable of providing; if women cannot bring down large prey, for instance, they are unlikely to have episodic food bonanzas to distribute. To look at how food is provided to family members compared to the group at large, Rebecca Bird, along with Brian Codding and Douglas Bird, reviewed foraging by men and women reported in three foraging peoples: the Aché, the Martu of northwestern Australia, and the Meriam from the eastern Torres Strait Islands.[25] These societies differ dramatically in their sexual division of labor, with men in the Aché contributing more than 85 percent of the food, while Martu women bring most of the calories to their groups. The researchers reasoned that at times when the available food sources were high-risk sources, women's contributions should be more important, since hungry children can't just wait for daddy to try again next week to bring down a kangaroo, and providing for children takes precedence over sharing with the group. Alternatively, when getting food is more reliable, men are expected to contribute more and emphasize sharing.

The researchers' predictions were upheld across the three very different societies, with the emphasis on both hunting and gathering changing depending on what was available to eat. For example, among the Meriam, turtle hunting is a chancy business much of the year, but during the nesting season it becomes more reliable. At that time, but not otherwise, the women participate by helping to plan the hunt and butcher the meat; unmarried men are the ones to take on the more risky hunting outside the nesting season. In all three societies, both men and women alter what they do depending on the likely yield of the catch.

The discovery that hunted food items, particularly big-game animals, are often shared among the group members, rather than being consumed only by the hunter's family, as well as the unreliable nature of hunted meat as a staple in the diet, has caused some anthropologists to question whether the main function of hunting is even subsistence at all. Kristen Hawkes, an anthropologist at the University of Utah, has suggested, instead, that hunting is a way for men to show off to prospective mates, and that good hunters gain high status in their social groups.[26] It is not that hunted meat isn't eaten and appreciated by the group, but that a given man's share does not necessarily increase the survival of his own children more than other sources of food do.

The unreliability of big-game hunting therefore may make it a poor choice as the sole source of support for a family; saying you want to maintain your wife and children on it is the ancestral equivalent of claiming that you will be able to fulfill your familial responsibility on the proceeds of playing lead guitar in a band. But hunted meat is not simply food—it is a signal. As Hawkes and Bird put it, "More than its value as a source of nutrition, meat is a medium of communication through which the hunter transmits information to potential mates, allies, and competitors."[27] By contributing to the group, good hunters gain the respect of their peers. The anthropologists point out that "showing off" in this context does not have to involve actual bragging; in fact, among the Aché, meat is often brought into camp quietly and without much fanfare. But everyone knows who can, quite literally, deliver the goods.

This "status signaling" hypothesis has been critiqued by other anthropologists, who continue to discuss how best to calculate hunters' contributions to their families, whether shared meat is repaid in kind, and which other factors play into the sexual division of labor. But the present-day perspective on hunting and gathering in modern—and presumably ancestral—societies still leads to three conclusions. First, sexual divisions of labor are widespread today, and very likely occurred in ancient humans, with men probably doing the more high-risk, potentially low-yield part of foraging. But second, and maybe more important, the tasks that men and women do are remarkably flexible across societies and over time. Anthropologist Jane Lancaster has noted that in various human cultures, women have been known to do virtually everything that men do, with a notable exception of metalwork, which she speculates requires too much single-minded concentration to be compatible with the presence of small children.[28] Finally, nothing about these sex-specific tasks, whether gathering limpets, digging yams, or snaring monitor lizards, suggests a justification for all manner of modern gender stereotypes, from women liking to shop while men watch football to an ability (or lack thereof) to ask for directions. That paleofantasy of the cavewoman staying home with the kids while the caveman went out for meat would have ended up with no one getting enough to eat.

Our bodies, our genitals, ourselves

Our behavior is slippery stuff, with men and women acting differently in different societies and under different circumstances. And without preserved footage of Neandertal dating sites, we are left to speculate about the sexual proclivities of our ancestors. Or are we? Behavior may not fossilize, but bodies do, and we can infer a surprising amount about behavior from those remains. For instance, we are reasonably sure that some dinosaurs cared for their young, because of the discoveries of eggs or young dinosaurs associated with an adult.

We can also draw conclusions about how natural and sexual selection have acted on the sexes in the past, simply by looking at the kind and magnitude of differences between male and female bodies today, as well as by comparing those differences across species. Along with the rest of his focus on sex, Darwin was extremely interested in these differences, and he drew a distinction between what he called primary and secondary sexual characteristics. The primary sexual characteristics are what define us as male or female—the plumbing, so to speak, with human (and other mammal) males having testes and females having ovaries, for example. All animals have primary sexual characteristics, and they can be very obvious or quite subtle; in many rodents, for instance, males and females are difficult to tell apart even after inspection of the nether regions, at least by a nonexpert.

The secondary sexual characteristics are even more variable, and they were also of much interest to Darwin. These are all the other differences between the sexes—the ones that aren't directly required for reproduction but are still sex-specific, such as the peacock tail I mentioned earlier or the songs of male frogs, crickets, or birds. Human secondary sexual characteristics include enlarged breasts in women, facial hair in men, and—most germane to our quest for the ancestral mating system—differences in body size, with men being on average 15 centimeters (about 6 inches) taller than women.

This body size difference is mirrored to greater or lesser degrees in many other species, though in some, including whales and many insects, it is the female that is the larger sex. (Larger females are believed to be favored when they can lay more eggs or otherwise provide more for the offspring.) Larger males are thought to be the result of sexual selection for better fighters, and the fighting is generally over access to mates. In elephant seals, for example, the 2-ton males are more than twice the size of the females, and the bulls spend hours battling for supremacy on the coastal breeding grounds. The champions are able to sequester and mate with a majority of the females that arrive on the beach, making the largest males big winners from an evolutionary perspective. So the general idea is that species with more male competition, and more polygyny, are likely to show a greater difference in body size between the sexes.

Among our primate relatives, the gorillas, which live in groups with a single male that mates with several females, have the most pronounced sexual size difference, with males about twice the size of females; similar sex differences are seen in orangutans. Males and females in the monogamous gibbons, by contrast, are roughly equal in size, as are the two sexes in the muriqui monkeys of Brazil, which have a polygamous mating system in which both males and females have several sexual partners during a season. Many primates also show sexual differences in their teeth, with male baboons, for example, having elongated, sharp canines that females lack. This difference, too, is thought to result from sexual selection, with the teeth used as weapons during fights between males.

Humans, along with chimps and bonobos, have a much more modest difference, which has led many researchers to conclude that we were only moderately to slightly polygynous in our evolutionary history. Fossil evidence also indicates that human sex differences in size have decreased over the last several hundred thousand years, though it can be difficult to draw conclusions, because skeletal remains are sometimes classified as male or female in the first place by comparing the size of the bones. Owen Lovejoy, an anthropologist at Kent State University in Ohio, extended this idea to suggest that human monogamy arose at least 4.5 million years ago, when the bipedal human ancestor *Ardipithecus ramidus* lived, and furthermore that it was facilitated by bipedalism, with men making use of the freeing up of their arms to hold tools, going off to hunt and bringing back meat for the women. Women, in turn, would reward their mates for the food with fidelity, à la Hrdy's "sex contract," tying things up in a nice tidy package.[29]

Of course, as I have already pointed out, this sharp division of labor is not upheld in modern hunter-gatherer societies, and furthermore, the similarity in relative sexual size between us and the chimps and bonobos suggests that we need more information before we can explain why we would end up monogamous and they would mate with many partners. Lovejoy also suggests that *Ardipithecus* was relatively peaceful, like the bonobos, but as de Waal notes, "unless the diggers come up with a male and female fossil holding hands and having wedding rings, the idea that these ancestors avoided conflict through pair-bonding remains pure speculation."[30] (One could nitpick that even this finding would not be conclusive, but de Waal's point is well taken.)

Furthermore, although very highly sexually dimorphic species are thought to be polygynous ("dimorphic" is the technical term indicating size and shape differences), it's not so clear what it means when the sexes are quite similar. Sometimes being bigger isn't all that useful, as in hummingbirds or other species with aerial displays of aggression, where a more agile opponent wins the day. Other variables, such as the likelihood of mating within one's group or outside it, can also influence the strength of selection on male competitiveness.

Even when we find evidence of sexual size differences, the degree of those differences is not necessarily the result of mating competition. Modern cultures vary in how dimorphic they are in height, with the Maya Indians of South America differing by nearly 10 percent, while the Taiwanese differ by only 5.5 percent. Claire Holden and Ruth Mace from University College London looked at seventy-six populations around the world for which they could find information on height, degree of polygyny, type of subsistence (hunting or agriculture), and the sexual division of labor during a preindustrial period.[31] If polygyny has made us dimorphic as a species, then it seems reasonable, if we look within human cultures, that the more polygynous cultures should also be more sexually dimorphic.

The latter two variables were of interest because the researchers hypothesized that polygyny might not be the only thing affecting sexual size differences; perhaps the way that men and women live is also a factor. And indeed, the more women contributed to food production in a culture, the smaller the difference in height was between the sexes, perhaps because such contributions meant that the women had more control over food distribution. Whether or not a society was polygynous made no difference, although the authors caution that their sample might not have allowed detection of a contribution by the mating system. They had to account for the historical relationships among the societies, as Fortunato did in her marriage analysis, described earlier; and once they did, they were left with a rather small sample.

The last piece of evidence about our mating history that we can glean from our bodies is a bit more personal than height. In many mammals, the size of the male testis is correlated with the number of females a male might potentially mate with over a short period of time. The testes, of course, produce sperm, and generally speaking, the larger the testis, the more sperm a male can manufacture. Usually, male animals produce enormous numbers of sperm cells (human ejaculates are 1.5–5.0 milliliters and contain anywhere from 20–150 million sperm per milliliter), but ejaculates need to be replenished, and might not be sufficient to fertilize the available females if a male is mating very frequently. In addition, if a female mates with more than one male in a short period of time, the sperm in her reproductive tract can compete with each other, in which case the male supplying the most competitors is at an advantage.

Several studies of animals, including primates, have found that species with more male competition for mates have larger testes than do species in which monogamy prevails. Interestingly, however, a 2010 study by Carl Soulsbury showed no relationship between the amount of mating outside a group—the "extra-pair paternity" that many paired-up animals show—and testes size among mammal species, suggesting that other factors besides sperm competition may be at play.[32] Soulsbury also found that species with larger litters had larger relative testes, again suggesting that male competition is not the whole story behind the evolution of our genitalia.

Where do humans fit into this picture? Human testes are smaller relative to body size than those of chimpanzees and bonobos, but larger than those of either the monogamous gibbons or the gorillas. Although male gorillas are substantially larger than females, because they live in groups where only one male routinely mates with the females, the silverbacks have relatively little need for competition with other males. Most researchers have concluded that this finding supports our decreasingly polygynous history, although Ryan and Jethá try to make the case that it points to a life of polyamory, with simultaneous multiple partners for both sexes.[33] They also note that men from different ethnic backgrounds differ in relative testis size, though since we do not know how each culture varied in the degree of multiple mating, it is hard to draw any conclusions from such information.

More recently, detailed examination of the stretches of DNA that are present in our ape relatives but absent in modern humans revealed a loss that women, at least, have cause to celebrate: the genes coding for "genital tubercles," or more graphically, penis spines. In many other mammals, including chimpanzees, the penis has hardened growths that may serve to sweep away the sperm of previous mates. These structures are absent in humans because we lack the genes responsible for the hormone signals that would cause them to develop. The relative smoothness of the human penis is thought to be linked to a reduced frequency of sperm competition.

So, is monogamy swimming against an evolutionarily promiscuous tide? I doubt it. Lifetime fidelity to a single partner may be rare among animals, and even among humans (whether it should be likened to corn syrup, as the quote at the beginning of the chapter says, I leave up to the reader), but the sheer variation in mating systems among human societies in both space and time makes it unlikely that we have all been ignoring our true natures. If evolution favored a single marriage or sexual system, why would we not all have converged on that pattern?

The piece of the puzzle that still needs addressing is the real evolutionary payoff: the "F" word. No, not that "F" word; several of the musings on human sexuality are as explicit as anyone could want. I am referring instead to "fitness," the biological term for success at passing on one's genes. A mating system will persist if its followers are better than others at having offspring that survive and reproduce themselves. How humans do that is the topic of the next chapter.

8

The Paleofantasy Family

Human babies are extremely peculiar little organisms. Children and the way they are raised are among the most unusual things about us. Our sex lives, while obviously of highlighted importance to us, are nevertheless in keeping with many other species: we have definite preferences for those we find attractive, we show off to prospective mates, and we compete with each other for access to those we favor. If you held your head the right way and squinted, we could practically pass for marmosets. Well, the lack of fur and the use of indoor plumbing would give us away, but you get the point. Monogamy is rare among animals, as we have seen, but hardly unheard of.

In contrast, no other species—primate or not—has such disproportionately large, demanding, slow-growing offspring. A comment on Paleohacks.com declares, "Human babies are wimps! They take FOREVER to become independent too! There's no other animal that takes anywhere CLOSE to the same length of time to start feeding themselves. What's up with that?"[1]

What's up with it indeed? Understanding how our slow development came about, and how our child-rearing practices do and don't reflect our evolutionary history, goes far toward explaining what makes us human—sex lives and all. What's more, sex means nothing in an evolutionary sense without, as I mentioned in Chapter 7, fitness, the bequeathing of our genes to succeeding generations. And fitness for humans means babies. If a practice enables our babies to be born and survive, evolution favors it. If a practice feels great, or even enables the practitioner to live longer, it doesn't matter from the standpoint of evolution if the long life or enhanced well-being is not translated into offspring, the currency of the biological realm. (Appearances to the contrary, children are like money.)

Let's examine the anomaly that is the human child. First, a comparison with our primate relatives. Anyone who has watched a mother chimp or gorilla at the zoo has seen the casual confidence with which infants are hauled around, their tiny faces bobbing behind or beneath the female as they cling to her fur. The mother does not have to work to keep the baby attached; as Sarah Hrdy notes, "Minutes after birth, possibly while the mother is still consuming the placenta, the tiny spidery newborn ape on the ground beside her will catch hold of her hairy belly and pull himself aboard."[2] Baby monkeys and apes are born able to do things that human infants can only dream of doing—assuming, that is, that their nervous systems were advanced enough to allow them to imagine an alternative, which they are not. Numerous researchers have suggested that human infants are in effect always born prematurely, with the nine months *in utero* only the beginning and an additional nine to twelve months required for the infant to reach a stage comparable to that of most other primates at birth.

Such prematurity is thought to be a necessary consequence of our large brains. A human infant is about 6 percent of its mother's body weight, while a baby chimpanzee is only 3 percent, although the adult females in both species are not so very different in size. Jeremy DeSilva of Boston University calculated the ratio of mother to infant body size in a number of fossil hominins, as well as apes, and suggested in a 2011 paper that our hefty babies evolved with *Australopithecus*, over 3 million years ago, considerably earlier than other anthropologists had believed.[3] DeSilva notes that lugging around such large infants would have been difficult for those early ancestors, lacking strollers and Snuglis as they did, which also means that *Australopithecus* would have had to give up life in the trees for a more terrestrial existence—a transition that again was thought to have occurred later in our evolution. Interestingly, anthropologist Tim Taylor suggests that the baby sling might have been invented by early humans, as a way to free the hands of bipedal animals, which in turn would have allowed the evolution of ever more large-brained and immobile infants.[4] This hypothesis is difficult to test, however, since the fibers that would have been used for such slings would not persist as archaeological artifacts.

That relative immaturity of our young persists into childhood. It is true that other apes nurse their offspring for longer than humans do, even in less westernized societies; chimpanzee infants, for example, wean at about four and a half years, while in many cultures children stop nursing after two to three years. But ceasing breast-feeding by no means signals independence in humans as it does in the other primates. Although they stop using their mother as an exclusive food source, young humans proceed to lollygag about before reaching sexual maturity. Other primates get on with their own reproduction relatively soon after weaning, with gorillas having their first baby just seven or so years later.

In effect, human beings invented childhood itself, an extended betwixt-and-between stage when a mother is less essential than in infancy but the youngster still cannot survive on its own. In contrast, a chimp orphaned or abandoned right after weaning at least has a fighting chance, as studies of wild primate populations show. And beyond childhood in humans is adolescence, another developmental stage virtually unknown among other mammals—occurring after sexual maturity but before having children of one's own.

Play between adolescent chimp male Titan and little Gimli. Individuals besides the mother often interact with youngsters. (Courtesy of Michael Wilson)

Even in nonindustrialized societies, girls do not start having children the moment they reach menarche; the average age worldwide for a first baby is nineteen, according to a 2008 article in *Science*.[5]

At the same time, ceasing nursing at a relatively early stage means that human mothers can have children more frequently than other primates can. Our "interbirth interval," the formal term for the time between the births of offspring, averages just over three years, whereas gorillas take nearly four years and orangutans a whopping nine. This ability to reproduce quickly is part of why our species has been so successful; in the same amount of time, and with the same starting population size, you can make a lot more people than orangutans. And those children are more likely to survive to adulthood, again even discounting the influence of modern medicine and hygiene, probably because of that long period of dependence.

What, then, is that lengthy period between weaning and adulthood for? And what does it tell us about the evolution of the human family, and about our paleofantasies?

Successfully failing to launch

Anthropologists often invoke one unusual human attribute when trying to explain another. For example, our extraordinary capacity for complex communication is sometimes linked to our complicated social structure, or our tool use to our large brains. When it comes to childhood, therefore, it was reasonable to suggest that a prolonged period before independence was required once humans began to perform difficult tasks, like hunting or making pottery and baskets. Children could spend their time practicing these skills, which would better prepare them for success as adults in a hunter-gatherer society. In effect, this idea would mean that children are schooling themselves, and were doing so long before formal education was invented. Many childhood games seem to be dress rehearsals for adult activities, and for our ancestors, instead of playing doctor, teacher, or princess, they might have practiced tracking prey or finding edible plants. (This leaves aside the question of just what those princess mimics are hoping to achieve, since presumably most of the Cinderella wannabes lack even a trace of royal blood, but that is an issue for another day.)

The problem is that, reasonable though this theory sounds, little evidence exists to support it. In at least one modern-day hunter-gatherer society—the Hadza people of Tanzania—some children grow up in the bush while others attend boarding school, where they learn to read and write but have little opportunity to acquire skills such as bringing down animals with a bow and arrow. A 2002 publication by anthropologists Nicholas Blurton-Jones of UCLA and Frank Marlowe of Harvard compared the skills of the two groups, as well as those of women, who grow up digging tubers in anticipation of their adult role, with those of men, who do not.[6] They paid the Hadza to participate in a kind of hunter-gatherer Olympics, with contests in various skills.

It turned out that the men were just as efficient at digging as the women, despite their lack of experience. And climbing baobab trees, "an important and dangerous skill" according to the anthropologists, was performed no better by young men who had been practicing during childhood than by those who were novices. It isn't that hunting and gathering aren't difficult, but that the critical time period for learning these skills may not coincide with childhood at all; the most effective Hadza hunters were around forty years old, suggesting that men needed to practice not just while playing as children, but as adults.

Blurton-Jones and Marlowe do not completely dismiss the idea of childhood as rehearsal, noting that skills other than those they measured may require key practice during maturation. But, they caution, "it is not safe to assume that increases in skill with age are entirely due to learning or practice; they may instead be due to increases in size and strength."[7] A similar comparison of adults and children of the Meriam people from the Torres Strait, Australia, also showed that the two age groups had comparable skills at fishing both with lines and with spears, despite the difficulty of these tasks.

Perhaps, then, the skills that must be learned during childhood are social, not physical. Given our complicated and subtle interactions, and the cutthroat nature of politics even in the smallest social groups, this is a reasonable supposition. But it, too, has had doubt cast on it; as anthropologist Meredith Small from Cornell University puts it, "Recent research suggests that children know more about social relations than we give them credit for, and they learn the ropes early."[8] Joking about schoolyard politics aside, even toddlers can resolve conflicts on their own, and eight-year-olds have social skills that enable them to see things from another's perspective and anticipate someone else's intentions. Little support exists for the idea that children must spend a long time as social apprentices before they can be successful members of a group.

What if, then, childhood evolved not for the sake of the children themselves, but for their parents? Anthropologist Barry Bogin suggests a rather cold-blooded analysis along these lines, in which, as he puts it, "childhood may be better viewed as a feeding and reproductive adaptation for the parents of the child" than as a period of coddled maturation.[9] Children in many societies perform useful work, ranging from care of crops or animals to help with their younger brothers and sisters, and they usually receive no compensation beyond their keep; allowances are a modern invention. In an even more Dickensian spirit, Bogin points out that "children are relatively inexpensive to feed."[10] Youngsters past weaning age thus can give back labor to help offset what they consume, enabling humans to reproduce at our uniquely rapid clip.

Meerkats, mothers, and sharing apes

In addition to having offspring with oddly prolonged childhoods, the human family is noteworthy for another unusual attribute: the common presence of multiple caregivers in addition to the mother. Human babies, like other infant primates, seek close and near-constant contact with their mothers. When anthropologists began studying maternal behavior in hunter-gatherer societies, they emphasized the way in which foraging women kept their babies close to them while they were digging for roots or picking berries. They were struck by how similar these women with infants slung on their backs seemed to be to any monkey or ape mother. This natural-seeming connection, apparently universal among our relatives, is part of what led to the attachment theory of parenting, in which the bonding between mother and infant is said to be an essential part of child development.

What the anthropologists did not notice, though, is potentially more significant than what they did. Like baby baboons and gorillas, human infants in contemporary hunter-gatherer societies are usually held by an adult, and babies spend little or no time entertaining themselves in the desert or forest equivalent of a playpen. The crucial difference is who's doing the holding. Among nonhumans, that individual is the mother most of the time, except for brief periods in which other troop members examine a newborn addition to the group. Even this limited sharing of infants varies considerably among different monkey and ape species, with langurs (Old World monkeys native to India) passing their babies to a host of aunts, cousins, and siblings almost immediately after birth. Red colobus monkey mothers are far more reluctant and will not allow their babies to even approach another female until they are three or four months old. Marmosets and tamarins, diminutive New World monkeys, are exceptions, with the father taking on the majority of child care virtually from birth—an unusual circumstance that I will discuss further later in the chapter.

In hunter-gatherer societies, by contrast, babies spend a substantial amount of time being held and cared for by someone besides their mother. They are certainly bonded to their mother and can recognize her from an early age, but they are carried around, fed (sometimes including nursing), and entertained by a variety of friends and relatives in their social group. These "alloparents"—people other than the mother who participate in child rearing—are thought to be the reason that humans can get away with weaning early and producing numerous children: exclusively maternal care is not the way that humans seem to have evolved.

In fact, when scrutinized closely, humans begin to look a bit like cooperative breeders. Say the phrase "cooperative breeder" to biologists, and their eyes light up in recognition. The same is not so true for most other people, so allow me to explain. Most animals have no parental care; the vast majority of invertebrates, as well as most fish, reptiles, and amphibians, unceremoniously dump their eggs or larvae into the environment and swim or waddle away. Even if they carefully craft a nest to safeguard the developing young, as do sea turtles, the adults are long gone by the time the babies appear.

In a number of species, however, someone, usually the mother, stays with the young until sometime after hatching. Birds feed their chicks, either by diligently ferrying worms to the nest as robins do, or by leading their young to food they can find themselves, as chickens do. In mammals, of course, maternal care is *de rigueur*, because by definition, mammals feed their young milk from the mother's mammary glands. In both groups, sometimes dad helps out, sometimes not.

Some birds and mammals and a smattering of other species go even further, and have individuals other than the parents helping to take care of the offspring. Because the behavior was first noticed in birds, it was initially called "helping at the nest," but once scientists found similar group rearing of young in mammals and even some fish, the more general term "cooperative breeding" became more common. In most cases, the helpers are older brothers and/or sisters who stay with their parents rather than going off to find a territory and raise a family of their own, although unrelated helpers are also observed.

Meerkats, those adorable little African animals from *The Lion King* and *Meerkat Manor* fame, provide a good example of cooperative breeding. Although they live in groups of twenty to thirty, only one pair in each group reproduces; the other members help with hunting insect prey, guarding the colony from predators, and taking care of the dominant female's young. The latter responsibility is taken to an extreme, with the females who lack their own babies producing milk for the dominant female's pups, as well as babysitting them. The African deserts in which meerkats live are hostile places, filled with snakes, scorpions, and other threats, and the abundance of caregivers helps ensure the survival of the pups.

The meerkat cooperative breeding system is fascinating for a number of reasons—how, for example, does the alpha female control breeding by the other females?—but for our purposes here, they are poster children for an alternative version of the family. Instead of mother and father and baby making three—or, given the usual meerkat litter size, making five or six—all the individuals in the group collaborate to raise one set of offspring. And in most cooperative breeders, including the meerkats, cooperating is the only way to go. A solitary pair of animals has very little chance of successfully raising a baby.

So, did humans evolve to be like meerkats, or more like the colobus monkeys, with their reluctance at allowing even other group members to handle their young? For many years, anthropologists and sociologists viewed the nuclear family as an essential element in human evolution, and as I discussed in Chapter 7, it was seen as the essence of our sexual behavior as well. And as far as child rearing was concerned, the mother provided all the necessary care, so long as someone went out and killed the mammoths. But increasingly, the idea that human mothers not only *can* share that care, but always have, has gained traction. In other words, humans act like cooperative breeders.

A few differences between humans and the classic vertebrate cooperative breeders are apparent, of course. Most important, in the meerkats and similarly behaving species, including birds such as the acorn woodpecker and the pied kingfisher, the helpers are forgoing their own reproduction by sticking around the familial territory. These additional individuals would be capable of breeding themselves, but for a variety of reasons they do not do so. In the meerkats and other cooperatively breeding mammals, the alpha female seems to influence the other females' physiology via chemicals she produces that temporarily suppress their sex hormones, rendering them unable to conceive. In some other cooperative breeders, the helpers could reproduce if they had their own nests, but acceptable real estate is in short supply, leading to a situation a bit like that of adult children returning to their parents' house after college.

In human hunter-gatherer societies, the helpers, or alloparents, are not quite so constrained; they can be siblings, grandparents, fathers, aunts, cousins, or even nonrelatives. How essential is their contribution to child rearing, and what does that level of importance suggest about the "natural" nuclear family?

Helping at the human nest

Anthropologist Sarah Blaffer Hrdy is well known for her studies of primate behavior, and while she is now most famous for her scholarly work on motherhood, she originally studied something far less beneficent: infanticide, the killing of young by other group members. In the hanuman langurs of India that she observed for many years, Hrdy discovered that a new dominant male entering the troop would try to kill the infants already present.[11] This horrific behavior is not an aberration, but an evolutionary strategy that is to the male's advantage, because a female who has lost her young will come into heat again very quickly, enabling the interloper to mate with her and increasing his fitness.

The female resists having her infant killed, but if it happens, her rapid return to sexual receptivity benefits not only the new male but her as well, since the resulting offspring will still pass on her genes. And it is this "whatever works" approach that puts the evolution of the family in a new light.

Hrdy agrees that human infants require a lot of care, and that early weaning combined with extended human childhood, even if youngsters contribute to the household, makes a mother's job a daunting one. A human mother can have one baby at her breast and several more that still depend on her—a situation that is difficult for one person to handle alone. Who helps out? Hrdy argues that although the "sex contract" view of human evolution that I described in Chapter 7—in which men bring home meat for the family in return for a promise of sexual fidelity—implies that the father is always the other component of the family, this need not be the case.[12] Why not simply use whoever is available, including, but not necessarily limited to, the father of the children? In other words, why not have a form of cooperative breeding?

Hrdy suggests that human females evolved to be opportunistic in their use of others as alloparents.[13] Sure, dad is fine, but if he dies, disappears, or simply can't come up with the meat that day, help from others should be perfectly acceptable, and in many parts of the world it is. Human mothers are likely to be the primary or only source of milk for very young children, but other forms of food can be provided by the rest of the group. Whether in hunter-gatherer cultures or modern industrialized societies, fathers are not reliably available, or they may not be up to the task. A harried mom seeking help from anyone who is available is not unusual or a product of a twenty-first-century life. "There is nothing evolutionarily out of the ordinary about mothers cutting corners or relying on shared care,"[14] says Hrdy. "Without alloparents, there never would have been a human species."[15] More globally, she points out, "The needs of children outstrip what most fathers are able or willing to provide," and hence "a mother giving birth to slow-maturing, costly young does so without being able to count on help from the father."[16]

Fathers can still be important, as I discuss later in this chapter. And Hrdy is not suggesting that we evolved in a chaotic environment of pass-the-baby, where everyone in a social group cares for the children *en masse*, in a kind of paleo kibbutz. Undeniably, human infants and their mothers form a close bond. But that bond is not, and maybe should not be, the only one that supplies the child's needs. Anthropologists Ann Cale Kruger and Mel Konner put it this way: "It may not take a village to soothe a crying baby, but it often involves more than a mother alone, a fact that is taking on growing importance in our understanding of human evolution."[17]

Studies of children reared with multiple caregivers—again, not in an atmosphere where babysitters come and go, but one with two parents and a grandmother, or with a mother, a couple of aunts, and an older sibling—support this idea. Among the Gusii agricultural people in Kenya, children are better adjusted, more empathetic, and more independent if they had strong attachments to at least one other person besides a parent while they were growing up. Anthropologist Karen Kramer of Harvard University calculated the proportion of direct child care, defined as nursing, feeding, carrying, holding, or grooming (this last refers to activities like bathing and dressing, not fashion or makeup advice) provided by mothers, fathers, and other individuals in nine traditional societies. On average, mothers did about half the work, with the remainder parceled out among an array of relatives and unrelated group members.[18] Hrdy summarizes a series of studies of Israeli and Dutch children raised either mainly by their mothers or by mothers and other adults by saying, "Children seemed to do best when they have three secure relationships—that is, three relationships that send the clear message 'You will be cared for no matter what.'"[19] Not one, not even two, but three. What's more, it is the reassurance itself that is important, not the person or persons who provide it.

Mothers, of course, often benefit from multiple caregivers themselves, in the form of increased child survival. An examination of eight traditional and modern societies by Ruth Mace and Rebecca Sear showed that having grandmothers who helped care for their grandchildren nearly always meant a greater likelihood of the children living to adulthood.[20] Similarly, Kramer found that the presence of alloparents in at least a dozen traditional societies meant better survival and growth of the children, as well as mothers who gave birth more frequently.[21] Kramer cautions, however, that the effect of these helpers is complicated by sharing of food and labor that goes on in the social group, apart from caring for the child per se; unlike the meerkats, of course, humans do not have a single breeding pair in isolation, so it is more difficult to separate the effects of alloparents' care from the general benefits of living in a social group.

A somewhat more skeptical view is taken by Beverly Strassmann of the University of Michigan, who has studied the Dogon people of Mali for over twenty-five years. The Dogon farm pearl millet and live in cliff-side villages with a close network of relatives. Although most men have a single wife, polygyny is accepted, and women have an average of ten surviving children. Most of the child care is performed by women and girls, with girls between the ages of five and nine often caring for their siblings.

Strassmann looked at child mortality, as did Kramer, but reported a very different picture. Among the Dogon, being in a kin group that extended beyond the nuclear family meant that a child was less likely to survive, not more—a pattern Strassmann attributed to increased competition for food and other resources.[22] In families with multiple wives, children grew more slowly and were more likely to die than in monogamous households. The grandmothers that were so helpful in Mace and Sears's study were viewed by the Dogon as a drain on society, although older women actually worked harder than older men.

Strassmann suggests that these dire findings result from parents and other relatives coercing children to work, so that what might superficially appear to be cooperation is actually forced labor. In addition, the aforementioned competition among siblings for limited food means that larger families will be harder on the smaller, weaker members. Finally, if men wield most or all of the political power, they can manipulate family life to their benefit and obviate any attempts at cooperation.

Calling all fathers, and grandmothers, and possibly aunts

To help resolve the disparity between the Dogon's situation and that of other traditional societies, as well as, perhaps, the ancestral human family, we need to look more closely at who exactly those alloparents might be. First, the most obvious supplement to a mother's care: the father. Although fathers cross-culturally tend to do little in the way of direct child care, exceptions do occur, most notably in the Aka foraging people of central Africa. Aka fathers spend more than half the day "within arm's reach of their one- to four-month-old babies," according to Hrdy,[23] and will often take children and even infants along on hunting expeditions. Significantly, hunter-gatherer fathers spend much more time with babies than do fathers from agricultural societies, with dads from Western societies somewhere in the middle. Perhaps the Dogon, with their agricultural lifestyle, lack the paternal help that some of the other groups with cooperative child rearing possessed.

Fathers sometimes get short shrift when it comes to involved parenting, but new research is suggesting that they might be underestimated. Conventional wisdom holds that men are unreliable long-term mates—and need to be bribed with sexual fidelity—because to them, mate competition is more important than child care. Hence, males were thought to benefit more by playing the field than by helping with the kids. But what if the urge to find a new mate is ameliorated by the experience of fatherhood itself?

This is more or less what anthropologist Lee Gettler and his colleagues found in a long-term study of over 600 men in the Philippines.[24] The scientists examined the men's levels of testosterone, the sex hormone behind many male-typical behaviors and physiological processes, before and after the men became "partnered fathers," as they put it. They also determined how much time each subject spent interacting with his child.

As Gettler and his coworkers predicted, men with higher testosterone levels at the start of the study were more likely to have become partnered fathers at the follow-up four and a half years later. But then something interesting happened. The fathers showed a dramatic decline in testosterone compared with both their own single, prepaternal levels and the levels of the men who had remained single. What's more, testosterone was lowest in those men who spent at least three hours a day caring for their son or daughter, after controlling for the effects of sleep loss and other variables.

This study is illuminating for several reasons. First, it uses longitudinal data, meaning that the same men were remeasured, allowing them to be used as their own controls. This kind of analysis is superior to one in which a group of fathers is compared with a group of single men, because a number of confounding factors could have caused a difference in hormone levels—say, if the single men were more likely to have used particular types of medication, or the fathers were from different socioeconomic classes. Second, the study indicates a finely tuned back-and-forth between a person's physiology and behavior. Cues from the environment can influence the hormone levels of fathers as well as mothers. The scientists suggest that although seeking a mate requires characteristics that may be antithetical to being a good father, it is, in fact, possible to have it all, and testosterone acts as the mediator. Third, as anthropologist Peter Gray pointed out in a commentary accompanying the article, the research "serves as a nice case study of the relevance of evolution to everyday human life."[25] The trade-off between mating and parenting is one that is predicted by evolutionary theory, and it means that a longing for new sexual partners might not be part of our heritage.

Gettler acknowledges that the degree of care by fathers is and always has been affected by many things, including geography, the way people make a living, and the importance of social class. He, along with other anthropologists, agrees that even when they do not hold or watch over children, fathers are often essential in providing food to their families. But Gettler is not willing to conscribe fathers to bread (or meat) winning alone. In a 2010 paper, he speculates that our image of early *Homo* society with a strict sexual division of labor may be incorrect, and that instead, "males, females, and young frequently traveled together to forage and scavenge at the same resource sites."[26] It is true that this pattern is not commonly seen among modern-day foragers, but if it occurs, it also solves another problem: who carries the baby?

As discussed already, being a hairless bipedal animal with a helpless infant is daunting, especially in an environment where you need to move around to make a living. One solution, as proposed earlier, is to construct slings or other technological gadgets to transport babies, but another, simpler, one is to have dad do the carrying. This solution not only frees mom's hands, but gives her a rest. From fossil remains, Gettler calculated the body size and the calories per hour that males and females of several early hominins would have expended in walking, and then extrapolated the increase in calories that would have been used (or saved) by carrying a 15-pound infant.[27] By having the father take over this task for some portion of the day, a mother could conserve her energy, and potentially be ready to conceive and bear another infant sooner, which would mean higher fitness for both parents.

Such optimistic scenarios notwithstanding (maybe those babywearing New Age dads are actually somewhat retro), fathers probably were not the sole sharers of child care in our evolution. For one thing, they can be unreliable, as mentioned already, both in the resources that they can provide and in the likelihood that they will stick around after a child is born. Jeffrey Winking and Michael Gurven examined the effect of fathers deserting their children to marry and have children with a new—and younger—wife on the survival of those original children, as well as on the fitness of the men themselves.[28] The researchers constructed a model using data on fertility rates and survival in four hunter-gatherer populations, as well as one agricultural group. They found that the men would not lose much in the way of child survival, and in fact had higher reproductive success overall, if they married a second wife who was just a couple of years younger than the first. Winking and Gurven's model makes a number of assumptions, such as that a new woman is probably waiting in the wings, but the bottom line is that selection may not have favored caring fathers simply to ensure the survival of their children.

If fathers are not the key, who is? A common answer has been grandmothers. Grandmothers are popular babysitters in many cultures—a choice that, besides being convenient, turns out to make good evolutionary sense. In the 1990s, anthropologist Kristen Hawkes proposed that maternal grandmothers were essential in our early evolution, because, being past their own reproductive years, they could help their daughters with the demands of young motherhood. In fact, she and others suggested, such duties could even explain an evolutionary puzzle: menopause, which otherwise seems like a peculiar attribute of humans. If having offspring is the way to evolutionary success, why would any organism live long past the time when it could reproduce?

At this point many people object that menopause would have been a nonissue in our early history, because the average life span was far less than it is today. While this is true, those averages can be deceiving, as I discussed in the Introduction; a high rate of infant mortality will make the average age at death low, even if a person who survives past twelve has a fairly high chance of making it to sixty-five. Indeed, a woman in a contemporary hunter-gatherer society who reaches menopause is very likely to live well past sixty, even without modern medicine.

According to what is called the grandmother hypothesis, these postmenopausal women were part of the alloparental team that enabled humans to keep our short interbirth interval and not have to rely on fathers for care. Having caregivers who did not have young children of their own would thus be favored by selection, leading to the evolution of menopause itself. Because of the lack of certainty that a given child is indeed genetically related to a given man, maternal, rather than paternal, grandmothers are thought to be particularly likely to fill this role. Caring for children pays off evolutionarily only if those children share your genes, and for fathers—and their mothers, the paternal grandmothers—some uncertainty of that relationship exists, even if the actual frequency of misassigned paternity is rather low. Hence, helpful maternal grandmothers are always contributing to the welfare of their genetic relatives, while paternal grandmothers may not be.

Proximity is another reason that grandmothers are on the list for helping out. In some societies, young women move to their husband's village or tribe when they marry, while in others the reverse occurs and women stay with their relatives while the men move. Hrdy argues that societies of the latter kind, which obviously favor maternal grandmothers as alloparents, were more common before industrialization.[29]

Whether grandmothers are truly linchpins of cooperative breeding, or their help is what drove the evolution of menopause, is still debated. Mace and Sear conclude that maternal grandmothers in their survey unequivocally helped their grandchildren survive,[30] and a study of modern Dutch families showed that when grandparents helped with child care, parents were more likely to go on to have more children.[31] The Dogon, of course, are evidence that grandparents are not uniformly helpful. As an alternative explanation, Michael Cant and Rufus Johnstone propose that, rather than grandmotherly help, the critical factor in menopause evolution was competition between the generations for resources to use in reproduction.[32] Edward Hagen and H. Clark Barrett found that among the Shuar people of Ecuador—horticulturalists who rely on plantain and sweet manioc—adolescent boys were the ones who helped children grow up, rather than their sisters or grandmothers.[33] Similarly, among two other hunter-gatherer societies in South America, men, not women, provided help to youngsters, albeit by giving them food rather than by cuddling them.[34]

Anthropologists will doubtless continue to debate the generality of grandmotherly help, along with cooperation in child rearing. A more important point, however, is that the image of the Pleistocene family as one of mother and father with their dependent offspring, operating largely by themselves, is inaccurate. And therefore the idea that we ourselves are best suited to a nuclear family life is misguided.

Crying, and sleeping, through the night

Thus far, I have been focusing on the evolution of children from the point of view of the parents, and assuming that once confronted with an infant, any mother—or alloparent—will know how to care for it. What about the baby's perspective? Advocates of "attachment parenting," "baby wearing," "cosleeping," skin-to-skin contact right after birth, and other supposedly new forms of parenting often suggest that these are more natural ways to care for children. "More natural" here presumably means that they are more likely to have been the ways that infants (and their parents) evolved.

Taking this idea to its more or less logical extension, some anthropologists have proposed the field of evolutionary pediatrics, which takes an evolutionary perspective to child raising. Just as evolutionary or Darwinian medicine suggests that we see increasing rates of "diseases of civilization" such as diabetes or hypertension because of the mismatch between our Paleolithic environment and our modern one, evolutionary pediatrics proposes that infants have the same needs that they had hundreds of thousands or millions of years ago, and that we have to find ways to satisfy these needs in an industrialized society. One champion of this approach, James McKenna from the University of Notre Dame, bluntly states, "We are pushing infant adaptability (and indeed maternal adaptability) too far, with deleterious consequences for short-term survival and long-term health."[35]

What does an evolutionary mismatch mean for a baby, who is not able to choose between high-fructose-corn-syrup-laden cookies and a more healthful option? Infants eat, sleep, and excrete, and it might seem that there is little that can be done to alter this pattern; many parents have marveled at the apparently primal needs of their newborns. Surely, if anything about human life " 'twas ever thus," including back in the Stone Age, it should be infancy.

One clue that this assumption might be false comes from a familiar source: cross-cultural data on parenting, including information about infancy in hunter-gatherer or horticultural peoples. Parents are usually quick to acknowledge that not all babies behave the same, but they don't necessarily recognize that babies from the same culture often have characteristic behaviors, behaviors that seem to arise from societal expectations. Cultural practices influence how, where, and when infants sleep, for example. We often imagine that babies move to their own rhythms, and that our ancestors must have responded to their infants in much the same way that we do. But this turns out to be incorrect; even among modern societies, some very basic aspects of infant life can be tremendously variable.

First off, sleeping. As the popularity of the not-for-children book *Go the F**k to Sleep*[36] attests, sleeping is a battleground for many parents; their children don't sleep long enough, won't go to sleep at the appropriate time, or cry inconsolably at intervals during the night. Is this natural?

Perhaps not. In Western industrialized societies, babies are encouraged to sleep in a way that they are not anywhere else—namely, alone in a room. In her fascinating book *Our Babies, Ourselves*, anthropologist Meredith Small points out that the Western emphasis on babies being independent from an early age is unheard of in most other cultures.[37] Babies are in near-constant physical contact with their mothers or an alloparent in virtually all foraging societies, where they can feed whenever they like; Small notes that the concept of nursing "on demand," or whenever the infant seems to want to be fed, as opposed to feeding on a fixed schedule, is foreign to these parents, since no one is demanding anything or acceding to demands. Small and McKenna also find baby monitors, devices that allow distant parents to hear their children crying or fretting, strange inventions; McKenna believes that babies evolved to sleep in social environments, and suggests that infants would be better served if the monitors conveyed the reassuring sounds of the parents' voices and activities to the baby, rather than vice versa. A study of parenting among the Maya of Latin America found that separating infants from their mothers to sleep was viewed as child abuse or neglect by Mayan mothers.[38]

Western parents often feel that a "good" baby is one who soon sleeps through the night, rather than waking frequently, but this ability may be common only among formula-fed infants and those who are placed on their stomachs to sleep. Sleeping on the stomach has been linked to sudden infant death syndrome (SIDS), and parents are now counseled to place babies on their backs, but the expectation of early independence at night is still widespread. Parents are also advised not to give in if their children wake frequently and demand attention.

We can trace the history of the Western concern that infants sleep alone back to a complex mixture of Freudian thought, concern about medieval mothers deliberately smothering their children, and other kinds of parental advice. McKenna and his colleagues examined the widespread belief that cosleeping—having an infant either in the same bed as the parent or in a bassinet or other device attached to the parental bed—is linked to SIDS, and found no support for the notion.[39] Instead, they advocate cosleeping as a means to strengthen the bond between mother and child and to ensure that the child breathes better and feeds appropriately. The environment, they say, should allow the infant and caregiver an opportunity to sense each other and continually adjust their positions and activities.

Interestingly, McKenna mentions at least one study examining the relationship between sleeping in a room away from the parents and later childhood behavior in which the children who had slept with their parents as infants actually demanded less, not more, of their parents' attention while playing.[40] In addition, Lee Gettler, mentioned earlier, coauthored a study with McKenna examining the sleep patterns of mothers and babies who slept in different rooms versus those who shared a surface at night or at least slept in the same room (cosleepers).[41] The two groups of mothers and their infants were then observed as they slept in a sleep laboratory for three nights. The experienced cosleepers breast-fed more often and tended to do so at shorter intervals—both attributes that are associated with optimal weight gain.

Crying is another infant quality that worries parents. Entire philosophies seem to be constructed around whether it is better to let children "cry it out" or to pick them up, and if so, after how long an interval, with concomitant worry over the eventual spoiling or emotional insecurity of the child. Here, Small flatly states, "There is extensive scientific evidence that the accepted Western caretaking style repeatedly, and perhaps dangerously, violates the adaptive system called crying that evolved to help babies communicate with adults."[42] She and other anthropologists and psychologists note that babies in non-Western cultures seem to cry far less than those from Western cultures. It's not that Western babies are simply unhappy. All babies seem to begin to fuss about the same number of times during a day or night, regardless of where they live. Scientists have even documented a so-called crying curve, in which babies seem to be hardwired to cry increasingly often until about two months of age, after which the amount of crying plateaus. But Western babies continue to cry in longer bouts, and their whimpering accelerates to real wailing more often, than do babies from non-Western cultures.

Kruger and Konner reviewed crying, and who responded to it, in groups of !Kung San people from Africa.[43] The !Kung babies cried for a maximum of one minute per hour (usually far less), and someone—usually, but by no means always, the mother—responded to 88 percent of all the crying bouts. Kruger and Konner characterize the society as "sensitive and indulgent" toward infants. They suggest that the !Kung's high responsiveness may make it less likely that babies escalate their crying to Western levels.

None of the researchers imply that Western parents are bad parents, that non-Western children live in an idyllic manner we should all emulate (for one thing, there is no single "non-Western society"), or that all babies that cry for longer than a minute are irrevocably scarred for life or have something wrong with them. All of the researchers stress the plasticity of human behavior, and the inevitable variation among infants and their parents that makes general proscriptions for child care risky. But babies do seem to have evolved in an atmosphere of immediate and frequent tending by multiple individuals, which suggests that large deviations from that environment may be hard on modern infants.

The findings also suggest that the baby wearers who promote carrying babies in slings or similar devices might not be too far off the mark. Of course, people will always interpret these suggestions from their own perspective; one website with many enthusiastic testimonials about slings had a comment about baby wearing in church.[44] The commenter was pleased that keeping her baby in a carrier had the benefits of "protecting your little one from germs" and "no playing pass the baby," noting with approval that "no one touched him"—needless to say, not exactly replicating hunter-gatherer life. I suppose you can lead a person to evolution but you cannot make her embrace its effects.

The conclusion that babies evolved with immediate care, and might not thrive if that attentiveness was missing, differs from accepting that babies are the way they are because they evolved in the Pleistocene that way, and that change since that time would be bad for them and for us—a claim that I find wanting. Instead of assuming that babies from foraging societies are more natural and more accurately reflect our evolutionary past, McKenna and others like him tested their ideas on real infants under modern-day circumstances. They found, for example, that babies cosleeping with their mothers breast-fed at shorter intervals than did babies sleeping on their own. We have obviously already altered child-rearing practices in many different ways around the world, and most children grow up just fine. So rather than concluding that we must adhere to a paleo way of child care, the question becomes which of those deviations is too great and which is an acceptable variant. The only way to address that question is with data. The way we think humans might have evolved is a starting point for asking questions such as, Does baby wearing keep infants from crying during the night? It is not, however, a prescription.

9

Paleofantasy, in Sickness and in Health

Let's face it—the reason people are trying to eat like cavemen or exercise barefoot is not because it looks (or tastes, or feels) good; it is so that they can live healthier, and preferably longer, lives. On the blog *Mark's Daily Apple*, for example, a commenter optimistically proclaims, "A diet that promotes high immune function—such as the high animal fat, low carb Paleo/traditional whole food diet—will protect against all cancers because it make [*sic*] the body strong."[1] Another contributor is even bolder, albeit with a rather wistful conclusion: "People need to realize they have total control of their health. Otherwise what kind of life is that destined to be sick by our own genes?"[2]

What kind of life is it, indeed? Do our genes, along with our modern lifestyles, inevitably make us sick? Or have the genes that can render us susceptible to illness also changed as we have evolved, making us better able to fight the diseases that always surround us? It turns out that some of the most exciting new developments in understanding human evolution come from medical research—studies not of cells in test tubes or mice in laboratory cages, but of the way our genes have responded to selection by pathogens.

Furthermore, one of the best places to detect the signature of recent evolution is in the diseases we get, mainly because it is easy to see the difference between winners and losers. After all, nothing says natural selection like a brisk round of the plague. But while the devastating effects of epidemics are well known, the accompanying evolutionary change that has occurred in human populations, some of it within the last few hundreds of generations, is not.

Understanding how resistance to disease evolves, and the exact nature of the genetic changes that enable that resistance, also has practical implications. From an evolutionary biology perspective, it is all very well and good to notice how gene pools change over centuries, but this information is not exactly helpful to medical researchers hoping to find cures within a patient's lifetime. If, however, we can determine exactly how the ability to withstand disease works by examining the recent changes in the DNA of the resistant individuals, we can try to mimic such alterations in developing treatments, without having to simply wait for a selective purge of more susceptible genes, and the people carrying them, from the population. And as I will detail later, that is exactly what scientists are doing in an attempt to treat AIDS, one of humanity's most recent plagues.

First, let's consider how the adoption of agriculture has affected disease, as it has affected so many other aspects of our lives. Much has been written about the changes in patterns of human ailments since the dawn of agriculture, most of it gloomy. As I mentioned in Chapter 2, becoming sedentary, working the soil, and keeping domestic animals all provide ample opportunities for the rise and spread of infectious agents like bacteria, viruses, and parasitic worms. Settled populations tend to be larger, making it easier to spread pathogens around; and keeping cattle, goats, or pigs means being in rather close proximity to their dung—a situation that both increases the diseases in the domesticated animals themselves and facilitates the sharing of parasites with them.

What's more, the influence of farming on the evolution of parasites was not a onetime event—our domesticated animals continue to evolve with their pathogens. A group of scientists from Norway and Switzerland led by Adèle Mennerat recently modeled the effects of intensive farming, defined as raising many animals in a small area with heavy management by people, on the evolution of parasite transmission.[3] Modern-day farming, whether of cows, chickens, or salmon, is a far different affair from the keeping of a few backyard animals; the creatures are kept in large numbers and are often cheek by jowl, in some cases literally, rather than in small flocks or herds. Such crowding selects for highly virulent parasites—those that do serious harm to their hosts—rather than for more benign forms that allow the host to survive and possibly recover.

These more deadly parasites evolve because, from the parasite's perspective, keeping a host alive longer is of little consequence, since another host is close at hand in the cramped quarters of the barn or fish hatchery. Likewise, such crowded conditions put parasites that develop more quickly and can move to a new host faster at an advantage; if the host population is spread out, a parasite that develops quickly might be too small or weak to make it to a new host. Such crowding is thought to have contributed to the high virulence of bird flu; the virus could easily move from one chicken or duck to another in the facilities housing many thousands of birds that were the norm in the parts of Asia where the virus developed. The poor hygiene and close quarters of the trenches of World War I may also have increased selection for the version of influenza that caused the deadly 1918 epidemic. Urbanization, with its dense populations all buying lattes or attending theater performances in the same places, is similar to both large-scale farming and the teeming hordes of war, at least in its ability to spread disease.

Mennerat and colleagues suggest that fish farming might be particularly vulnerable to such conditions, and they recommend that agricultural managers and evolutionary biologists collaborate to avoid the evolution of such deadly diseases.[4] For example, keeping smaller populations of fish in more ponds, rather than as a single megaschool, reduces the advantage for a parasite to become more virulent.

In addition to potentially fostering more harmful diseases, agriculture is sometimes thought to cause illnesses more directly, by producing less healthful food. If the grain-is-evil crowd is to be believed, our nutrition became worse once we stopped being hunter-gatherers, leading to increases in noninfectious ailments like diabetes and obesity that further compromise our health. In *Pandora's Seed*, Spencer Wells rather despairingly notes, "It's as though agriculture were a virus, expanding in influence despite its negative effects on human health."[5]

I have been emphasizing how much our genes, as well as our lives, have changed in the 10,000 years since agriculture, making us different in many ways from our Paleolithic ancestors. How many of those changes have involved our health, and were they indeed always negative, as the doomsayers claim? Furthermore, as new diseases emerge, will our genes be able to meet the challenge?

Everyone dies of something

It is obviously reasonable to assume that everyone who lived before us also died, and it's often possible to determine the approximate age at which those deaths occurred, either from remains or, for more recent cases, from records kept over the last few centuries. Less clear is the cause of each demise. Any attempt to determine the cause of death for ancient people is complicated by at least two issues. First, evidence of many diseases simply disappears as bodies decompose; mummies from ancient Egypt and the occasional "bog person" preserved in permafrost or peat can provide more information than skeletons, but these complete specimens are relatively few and far between. And bodies will not bear the evidence of infectious diseases such as cholera or influenza, because those diseases tend to kill too quickly to leave signs of their presence on the skeleton.

Second, using written records of the cause of death is tricky because our ability to diagnose illness has improved so much over time. Someone who was noted to have died of "fever" may have suffered from malaria, pneumonia, or sepsis—or none of these. Children were sometimes said to die of hives, a skin condition that accompanies a variety of diseases but that is rarely fatal in itself, or from teething; the latter may have been due to lancing swollen gums with dirty instruments, to subsequent infection, or to the teething child's being weaned and switched to fluids contaminated with pathogens.

With these caveats in mind, anthropologist Timothy Gage calculated mortality rates and causes of death for humans from different places and times using a variety of sources.[6] He focused on the last century and a half, but also included some prehistoric data. Although mortality rates were clearly higher in prehistoric times than they are now, contrary to what is suggested by Wells and others despairing about the effects of human settlement, people did not seem to die at an earlier age once agriculture was adopted. Note, too, that average life expectancy, as I discussed in the Introduction, is just that—an average; high childhood mortality from infectious diseases such as diarrhea can make it seem as if people in ancient times keeled over at age thirty-five, when in fact, if they made it past seven or eight their chances of living past sixty were reasonably good.

Gage suggests that the reason other studies found lower life expectancies for populations with agriculture than for those before settlement is that agricultural populations usually grow at a much more rapid rate than nonagricultural populations. This means a lot of babies being produced. More babies mean more opportunity to succumb to the aforementioned diseases. This high childhood death rate makes the average life expectancy seem lower than it actually is. And while we may bemoan the so-called diseases of affluence, Gage also points out that in most parts of the world, mortality rates have decreased substantially over the last 300 years.

Is that decline simply a result of fewer infectious diseases in modern populations, with degenerative ailments like cardiovascular disease still increasing as part of the price we pay for agriculture and sedentary living? Gage thinks not, for two reasons. First, some of the deaths may have been mistakenly attributed to degenerative diseases, when in fact they were due to other causes, from infections to accident. Second—well, we all have to die of something. Infectious diseases are declining, so other causes of death will take their place. What's more, Gage is skeptical of the reports of the low incidence of degenerative disease in contemporary hunter-gatherer societies. Virtually all such groups studied by anthropologists are very small, making the total number of deaths that would occur during a given field study similarly tiny and the drawing of conclusions about their cause risky at best.

Some solutions to the controversy may be forthcoming, as the results of the European Global History of Health Project become available. This ambitious effort brings together data on skeletal remains from thousands of years ago from across the continent, attempting to evaluate evidence of injury, joint damage from osteoarthritis, dental status (a good indicator of diet), and other health markers.[7] By pooling the efforts of seventy-two researchers, the project will be able to overcome earlier difficulties of small sample sizes from isolated locations. Initial results suggest that people living in early urban settlements were indeed of poorer health than their hunter-gatherer ancestors, but that health improved later, after trade networks allowed the exchange of goods and food became more diverse. Newly agricultural societies may suffer, but the steady source of food, and the potential to trade for the foods not locally available, seems to pay off later.

Sickness old, sickness new, and sickness borrowed too

We can blame animals for some of our diseases, but in a larger sense our own genes are also at fault. Diseases can be divided into several categories, and our genetic heritage—and evolution—can influence all of them, though some of the effects are more direct and others more subtle. First and most obvious are infectious diseases—those caused by living things: viruses, bacteria, or larger parasites such as worms. Here our genes act to make us more or less susceptible to infection; the immune system varies in its efficacy, and we can inherit a vulnerability to particular types of infections.

Second, some ailments are due to defects or vulnerabilities in our genes themselves. Huntington's disease, for example, is a neurological disorder that is caused by a defect on one of the chromosomes; people who suffer from the disease have a segment of DNA that is repeated too many times, and they will inevitably develop a variety of debilitating symptoms. Of course, even in genetically based diseases, the environment can play a role. Diabetes has a genetic component and runs in families, but its manifestation can be drastically influenced by the lifestyle of the person with that genetic predisposition.

Finally, degenerative diseases such as cancer or cardiovascular malfunction can occur because of a complex mixture of genetic and lifestyle factors. Some scientists have even suggested that infectious agents, especially viruses, can play a role in heart disease or other illnesses previously thought to be solely due to deterioration of the body's systems. Regardless of whether this idea pans out, it is clear that genes associated with disease are both important and numerous: from actual disease-causing genes, to genes that influence the operation of the immune system, to those that affect the strength of our tissues and hence their vulnerability to failure, much of our genome is occupied with keeping us healthy. Therefore, such genes are particularly good places to look for evidence of recent evolution.

Humans, of course, are not unique in our ability to get sick; not only do big fleas have little fleas upon them, as the doggerel goes, but animals, plants, and even bacteria are subject to various ailments. The genes associated with defense against those diseases are passed along when new species evolve. In trying, then, to understand how genes related to health evolved, we need to look at where they originated in our evolutionary history. Modern genomic techniques now allow scientists to compare the DNA sequences of different organisms and calculate the time at which various genes of two species, or higher-level groupings like birds and reptiles, diverged. So, for example, we can see which of our DNA sequences we retain from our common ancestry with fish, how many from our ancestry with our fellow primates, and so on.

Evolutionary geneticists Tomislave Domazet-Lošo and Diethard Tautz did just that, using an enormous database that catalogs human genes associated with diseases.[8] The database contains over 4,000 chromosome regions that are associated with a genetic disease, meaning that people who have the disease are more likely than the population at large to have that particular gene variant on their chromosomes. After "cleaning up" the data by removing questionable sequences or other potential errors, the researchers ended up with 1,760 disease genes based on Morbid Map, the delightfully named guide to the location on the chromosomes of genetic disorders that was developed at Johns Hopkins University. They then compared the genes to those of other creatures on Earth, from bacteria to primates.

Surprisingly, well over half of the human disease-related genes were very old, dating back to the origin of life itself. This is not simply because many of our genes themselves are ancient; 40 percent of our genes in total come from bacteria, but about 60 percent of the disease genes share their ancestry with bacteria. Less than half of 1 percent of the disease-related genes are so recent as to be shared only with the early mammals, which originated about 240 million years ago. The map does not allow assignment of particular ailments, but it can point to processes, such as the breakdown of food into usable elements within a cell, that are more likely to be associated with disease-related genes than with other types of genes.

Some of the media commentary on this work played up the potential practical application of our shared diseases, noting, as did Domazet-Lošo and Tautz, that the existence of the genes in simple organisms like worms, insects, or bacteria means we could study diseases in animals that are much easier to experimentally manipulate than the currently popular rats and mice. But Domazet-Lošo and Tautz also rather flatly note that "genetic diseases are an inescapable component of life."[9] Their results point to the deeply rooted nature of disease itself. In an ideal world, natural selection would weed out those who cannot resist disease, but the only world we have is not ideal. It is just the one we have.

We carry our susceptibility to diseases with us, embedded in our genetic core, unable to be shed when our ancestors separated from worms and turtles, far before the time when we came down from the trees, left the savanna, and started carrying spears and babies. At the same time, natural selection constantly favors genes that make us healthier, particularly in response to the challenges of new diseases. This continuity between ancient and modern is further argument against our suddenly having wrenched our Stone Age selves into a jarring new environment to which we are not adapted. We are both always facing new environments, and always shackled by genes from the past. After all, those Paleolithic ancestors were still dragging around genes they shared with hamsters and bacteria.

The genes fight back

Saying we still carry old disease-causing genes tells only part of the story. We may have inborn vulnerability to illness that we cannot discard, but selection has not been completely slacking off. In addition to those older genes, we show many new ones, some of which arose well after the dawn of agriculture. And understanding those changes is leading to treatments for the most current of plagues: AIDS.

A gene with the catchy name of *CCR5* codes for a protein on the surface of white blood cells known as T cells, critical components of the ability to recognize and fend off foreign invaders such as viruses. Some people have a variant of the gene called *CCR5*-delta (the "delta" is usually represented by the Greek letter Δ). If they have one copy of the variant, they are resistant to the human immunodeficiency virus (HIV), which causes AIDS. Such people still become HIV-positive, but the actual disease manifests itself two to three years later in them than it does in people who have a different form of *CCR5*. People with two copies of *CCR5*-Δ are essentially immune to the virus—an extraordinary characteristic, given the usual destructive nature of HIV infection. *CCR5* genes in their usual form allow viruses to penetrate the outer membrane of cells, but individuals with the *CCR5*-Δ gene derail that entry before HIV can get through.

In 1998, a group of scientists led by J. Claiborne Stephens and Michael Dean from the National Cancer Institute surveyed 4,166 people from thirty-eight ethnic groups across Eurasia, Africa, and East Asia, as well as some Native American populations, for the presence of the *CCR5*-Δ gene.[10] The proportion of each population that carried the variant ranged from zero in the East Asian, Native American, Middle Eastern, and Georgian groups to over 13 percent in Sweden, Russia, and Poland. Relatively few Greeks bore the gene, but it was more common among eastern Europeans. In general, the gene occurs at higher frequency in northern countries than in southern countries.

Using a mathematical modeling technique, the researchers were able to estimate when *CCR5-Δ* is likely to have originated. They determined that the gene variant is about 700 years old, though it could be as young as 275 or as old as 1,875 years. Why do we see variation in the proportion of people in the different populations? Presumably, the gene rose in abundance because it was favored by natural selection. From that average starting point of 700 years ago, the scientists calculated how much selection had to favor the gene in different geographic locations in order to produce the proportions we see today. It turned out that selection had to have been quite strong, meaning that people carrying *CCR5-Δ* would have to have outreproduced those with the more common version of the gene by quite a wide margin.

What advantage could those people have had? HIV itself came on the scene only within the last several decades, but the association between *CCR5* and the immune system suggests that disease played a key role in the selection process. The obvious candidate seemed to be bubonic plague (the Black Death), a bacterial disease spread by rats harboring plague-infested fleas. During the Middle Ages, bubonic plague killed between 25 and 40 percent of the population of Europe, with some of the largest epidemics occurring—you guessed it—about 700 years ago. Later epidemics, though still destructive, killed a smaller proportion of the population, and plague declined in Europe after a pandemic in 1665 and 1666 called the "Great Plague." The disease had all but disappeared from Europe by 1750. It would make sense for selection conferring resistance to bubonic plague to have driven the gene to high frequencies in a relatively short time.

This idea was at least tentatively accepted for several years following Stephens and colleagues' publication, though the authors themselves were cautious in necessarily making plague the culprit, or hero, depending on your point of view. They suggested that the bacteria *Shigella*, *Salmonella*, and *Mycobacterium tuberculosis*, which all cause human diseases, were also suitable candidates. But in 2003, Alison Galvani and Monty Slatkin from UC Berkeley published a paper expressing skepticism about the plague hypothesis.[11] Instead, they favored smallpox, another devastating disease, but one caused by a virus, not a bacterium, and one that is spread directly, from one person to another, rather than via a vector like the flea.

Plague and smallpox differ in many ways, but key to Galvani and Slatkin's argument is the way that these diseases move through populations. Plague roars through like a wildfire, slaughtering victims of all ages and then disappearing, at least for a while. Smallpox is more low-key; its direct transmission means that it tends to persist at lower levels among children and young people, who pass it among themselves but then either die or survive with lifelong immunity to future infections. Selection from smallpox is therefore more continuous, rather than the intense but periodic carnage of plague.

Galvani and Slatkin mathematically modeled whether a disease like smallpox or one like plague would be more likely to produce the frequencies of *CCR5*-Δ that we see today.[12] They included several assumptions about how often mortality from either disease reached its highest level, how much disease resistance the gene would confer, and, perhaps most significant, the age structure of the group, meaning how many younger versus older people are in a given population. This last assumption is important because selection acts differently on organisms, depending on how much, if at all, they have already reproduced. Selection for resistance to childhood diseases takes a while to exert its effect, because of course the benefit of being resistant—namely, staying alive long enough to reproduce—won't be seen until the resistant children reach sexual maturity. At the same time, such diseases can have a bigger effect on the population because they remove more "reproductive potential"—the modelers' term for one's promise of having babies. By contrast, someone who has a disease like plague that attacks all ages indiscriminately might already have donated plague-susceptible genes to the gene pool before being selected out of the population.

The calculations showed that bubonic plague was unlikely to have driven the increase in *CCR5-Δ* bearers that occurred between the thirteenth century and today. Smallpox, on the other hand, stays in the running. Although it did not wipe out as many people at a time as the plague, the total number of deaths caused by smallpox over the last 700 years is greater than those caused by plague, and it fits nicely into the other required criteria for causing the uptick in *CCR5-Δ*. Slow and steady wins the race, at least if by winning you mean putting more people into an early grave.

How did the *CCR5-Δ* gene variant spread across Europe and western Asia? Understanding the movement of the gene can help unravel the selection behind it. Along with John Novembre, Galvani and Slatkin tried to distinguish several possible ways the gene might have dispersed.[13] First, *CCR5-Δ* might have originated in northern Europe and been carried south, either through normal relatively slow migration or by the Vikings as they advanced through the continent in a more concentrated burst. Alternatively, the gene variant could have arisen in central Europe, becoming more common in the north because selection was stronger there. Yet another possibility is that although *CCR5-Δ* helps its bearers resist plague, it also has a cost in susceptibility to other diseases, and that cost was higher in southern regions, leading to a northern bias in the prevalence of the gene variant.

Again employing a series of sophisticated mathematical models (in another charming bit of jargon, they tried out a possible dispersal distribution called a "fat-tailed double exponential," which sounds to me more like a rare type of bird than a mathematical element), the researchers were able to conclude that the "Viking hypothesis," in which the Vikings took the gene with them as they moved through Europe, was consistent with the data. The gene variant probably arose either in Spain or northern Germany and spread outward from there before the Vikings moved it. The investigators also confirmed their earlier idea that selection on the gene must have been quite strong, meaning that the advantage it conferred was important enough to drive the rapid increase in the frequency of *CCR5-Δ*. That selection would have eventually increased *CCR5-Δ* in other populations besides the ones that exhibit it now, given enough time.

A final nail in the coffin of the Black Death hypothesis for *CCR5-Δ* (a little bit of plague humor there, if I may) comes from the tiny country of Malta, an island that was under Norman rule for over 400 years in the Middle Ages. Since the fourteenth century, Malta has suffered three major outbreaks of bubonic plague and several more minor ones. The Vikings or other Europeans would have had ample time to introduce the gene, and the epidemics would have placed its bearers at a considerable advantage. The Maltese, however, show virtually no evidence of *CCR5-Δ*, as a survey of 300 blood donors by Byron Baron and Pierre Schembri-Wismayer of the University of Malta determined, even though they would have been expected to be prime candidates for selection.[14]

Being careful what you wish for

At this point one might conclude that it is good to have the *CCR5-Δ* variation, and because such disease resistance can arise and spread rapidly in human populations, one could further hope that even though agriculture has introduced new ailments, we will simply be able to adapt our way out of these newfound threats. Not so fast. Evolution often works by robbing Peter to pay Paul, such that an advantage in one situation means a disadvantage somewhere else. This is not intentional, of course; fatalistic superstition to the contrary, it is not as if Mother Nature penalizes those with robust immunity to smallpox by making them susceptible to warts. Instead, it is just that change in one part of the genome can have consequences elsewhere. Note that I did not say "unforeseen" or "unintended" consequences; as I have mentioned previously, everything about evolution is unintentional.

In the case of *CCR5-Δ*, one of those consequences could lie in susceptibility to another disease: West Nile virus, a mosquito-borne disease originating in Africa and first discovered in the United States in 1999. Many people harbor the virus without knowing it, but those who are immunocompromised—say, from HIV infection—are prone to develop a more serious form of illness. Complications of the central nervous system, particularly encephalitis, can in some instances even lead to death. In 2011, 690 cases of West Nile virus were reported to the CDC, with forty-three deaths.[15]

Because of the immunity to AIDS that *CCR5-Δ* offers, medical researchers have attempted to mimic the effects of the gene variant in devising treatments for AIDS. If they can reproduce the change in the immune cells that *CCR5-Δ* confers, they should be able to reproduce the resistance as well. But in 2006, Philip Murphy and his colleagues from the National Institute of Allergy and Infectious Diseases discovered a snag.[16] Mice that were genetically engineered to have *CCR5-Δ* were unusually susceptible to West Nile virus. More troubling, symptomatic West Nile virus was much more common in blood samples from people with the *CCR5-Δ* variant than would be expected by chance. The mechanism for the heightened susceptibility is not yet clear, but in the meantime, altering *CCR5* as part of the treatment for AIDS or other diseases might not be a good idea, if it could have the side effect of increasing the risk of West Nile virus infection.

Such no-free-lunch findings are in keeping with what we know about adaptation to disease, or for that matter, adaptation to any other threat in nature. It is not a matter of all genes serving their own essential task, so that selection cannot interfere without throwing a monkey wrench into the DNA. Scientists have known for many years that the genome contains much duplication of function, not to mention what is sometimes called "junk DNA," genetic material that seems to float around in the chromosomes, shockingly unemployed. No modern geneticist would expect to see a system so tightly orchestrated that the slightest deviation would send the whole shebang into disarray.

Indeed, it is precisely because of that multiplicity of function that a new mutation in one area of the genome, rendering resistance to a particular disease, may well have effects on other areas. No one stood over our chromosomes assigning each gene a single task, like a worker on an assembly line. Instead, it is as if the worker making the green widgets also happens to be in charge of cleaning the bathrooms, not to mention staffing the reception desk. Changing the widget color that a person produces from green to purple might vastly increase widget sales, which is terrific for the company's bottom line, but it might also mean that phone calls go unanswered and toilets overflow.

Similar win-some-lose-some scenarios apply to the genetics of vulnerability to several other modern diseases. Cystic fibrosis, for example, is the most common fatal inherited disorder in populations of European origin. People with the disease have thickened mucus in their respiratory and digestive tracts, leading to a variety of complications. Those with one copy of the cystic fibrosis gene, however, do not show the disease; and recent research suggests that the gene may have persisted in human populations because it also confers some degree of resistance to cholera, another often-fatal bacterial disease, spread via contaminated water.

All of these trade-offs underscore the point that we did not evolve to be in perfect harmony with our environment, whether in the Pleistocene or otherwise. You win some (in the form of increased immunity to AIDS or the ability to withstand dehydration), you lose some (maybe via susceptibility to West Nile virus or in a nasty lung disease). What's more, 'twas ever thus, and cavemen were not any more likely to escape those balancing acts than we are.

Bred in the bone

One ailment has long been the poster disease for the woeful consequences of adopting agriculture: tuberculosis. Unlike many other infectious diseases, it leaves marks on the skeleton that allows its diagnosis in ancient remains, so it has provided fodder for theories about how health has worsened in recent times. A variety of irregularities, including spinal curvatures and bone destruction followed by new bone formation, are associated with chronic tuberculosis infection, and they can be used to infer the rates of the disease in ancient populations. In mummified bodies, like those from ancient Egypt, the lungs and other organs can also show signs of the disease.

Tuberculosis is caused by a bacterium called *Mycobacterium tuberculosis*, and because a similar type of bacterium, *Mycobacterium bovis*, is found in cattle, it was often assumed that the disease spread from cows to people after the domestication of animals became common. Further support for this idea came from the relative rarity of tuberculosis-infected remains from the period before people began to settle down.

More recent research, however, calls that scenario into question. In addition to simply examining ancient bones, scientists can now detect the bacteria themselves in specimens. They can even determine the precise strain of either *Mycobacterium tuberculosis* or *Mycobacterium bovis*, so that they can tell, for example, whether all members of a family harbor the same strain, suggesting they passed the disease among themselves; or they differ, suggesting that each person acquired his or her case independently. This technology has revealed tuberculosis infections in skeletons that showed no visual signs of disease, and it has suggested that tuberculosis might well have been common far earlier than was previously supposed.

New detailed studies of the differences between *Mycobacterium tuberculosis* and *Mycobacterium bovis* also cast doubt on the idea that the cattle form was the ancestor to the human disease; the DNA in the two types of bacteria is just too dissimilar to easily imagine one form evolving into the other in the time frame that would be required. Furthermore, if the cattle version had passed to early humans, one would expect to see *Mycobacterium bovis* in numerous ancient human populations; instead, however, the cattle form of the disease has been detected in only one small sample of Iron Age pastoralists from southern Siberia. The latest estimates now place the evolution of *Mycobacterium tuberculosis* at least at the time of the early hominins, between 2.6 and 2.8 million years ago, though these dates are still debated by scientists.

Further support for the idea of tuberculosis as an ancient human disease comes from Helen Donoghue of the Centre for Infectious Diseases and International Health at University College London, who suggested that tuberculosis might have coexisted with early humans while they lived in small hunter-gatherer groups.[17] If infected adults are healthy, the disease may cause relatively little damage, becoming virulent only if the person harboring the bacteria grows old and becomes immunocompromised or has other stressors and diseases at the same time. This idea, that tuberculosis can be a chronic but not bothersome infection unless the immune system is otherwise compromised, is also in keeping with the current concern about outbreaks of tuberculosis in HIV-positive populations, since they, too, have weakened immunity.

The tuberculosis-as-curse-of-modernity idea is not totally dead, however, and the latest twists in the story of this disease showcase the continuous nature of our evolution with, and against, sickness. Donoghue suggests that as humans lived in larger and larger groups, selection on the bacteria would favor strains of tuberculosis that spread quickly and were quite harmful to their hosts.[18] Just as in the farmed animals—or the city—if a new host lives in the same house, or the same neighborhood, as the current one, it is no longer beneficial to the pathogen to keep its old host alive for as long as possible, since a juicy new victim is literally just around the corner. Hence, a strain of tuberculosis that ruthlessly exploits its hosts will pass on more of its genes in a crowded environment, while a somewhat more restrained form will be more successful in a sparsely populated one, all else being equal. And indeed, the newest strains of *Mycobacterium tuberculosis* progress more quickly to full-blown disease.

Tuberculosis doesn't get the last laugh, though. A group of scientists from the United Kingdom and Sweden led by Ian Barnes looked for the frequency of a gene variant associated with resistance to tuberculosis in DNA samples from seventeen human populations across the globe that had been living under crowded urban conditions for differing lengths of time.[19] A population of Anatolian Turks, for example, had been living in settlements since 6000 BC, whereas a Sudanese group had been urbanized only since 1919. Barnes and his colleagues predicted that the longer the population had been living in closely settled conditions, the more common would be the resistance genes. They were right; the more urbanized groups had higher levels of resistance, with a response that evolved in a mere handful of generations.

Why doesn't everyone, or at least everyone in urban populations, show the same resistance? Once again, there is no disease-free lunch. People with the tuberculosis-resistant genotype appear to be more susceptible to autoimmune diseases, in which the immune system is so good that in effect it turns on itself, causing self-inflicted damage. Nevertheless, in this case—and probably many others—sticking to the genes of our ancestors, or assuming those genes were better adapted to the environment, would have been a bad idea.

Cancer: Old enemy or newfangled foe?

Cancer is a frightening disease, and most of us have either experienced it firsthand or had a close friend or relative diagnosed with the disease. It is the second leading cause of death in the United States, according to the American Cancer Society, which further notes that as of 2011, half of all men and a third of all women in the United States will develop cancer during their lifetimes.[20] At the same time, perhaps because of all the high-tech detection and treatment involved in cancer, it often seems as if the prevalence of the disease is a recent phenomenon, and that back in the day people just didn't suffer from cancer the way they do now. A 2007 survey, also from the American Cancer Society, found that nearly 70 percent of Americans believed the risk of dying from cancer to be increasing in the United States.[21] Along these lines, early anthropologists living with foraging peoples have sometimes remarked on how healthy and apparently cancer-free their subjects' lives seemed to be.

At least some of the paleo diet and lifestyle enthusiasts place the blame squarely on, you guessed it, the postagriculture environment and its diet of grains and other processed foods. This idea is not new; in 1843, French physician Stanislaw Tanchou gave a presentation to the Paris Medical Society arguing that grain intake was a strong predictor of cancer incidence.[22] He also predicted that cancer would not occur in hunter-gatherer populations—a notion that, according to the website NewTreatments.org, was born out by "a search among the populations of hunter-gatherers known to missionary doctors and explorers. This search continued until WWII when the last wild humans were 'civilized' in the Arctic and Australia. No cases of cancer were ever found within these populations, although after they adopted the diet of civilization, it became common."[23]

The actual data from Tanchou's research seem to be unavailable, and it's hard to know how far to trust a website that includes the statement, "The author and publisher can't be held responsible for anything."[24] Nevertheless, the question raised is legitimate: Is cancer one of the manifestations of our deviation from the environment for which we are best suited? Or has it always been with us? Scientists have been divided on the subject, with some medical researchers noting the types of cancers attributable to modern environmental influences such as pollutants, and others claiming that improved diagnosis makes it seem as if the disease has become more common.

Interest in the idea of cancer as a curse of civilization, or at least of modern urban living, was renewed in 2010, when Egyptologists Rosalie David and Michael Zimmerman published a review of literature on cancer in prehistoric peoples, including surveys of mummified remains, as well as ancient writings.[25] They detected remarkably few instances of cancer, leading them to conclude that "cancer was rare in antiquity." In their paper, David and Zimmerman did not implicate diet in the apparent rise in cancer rates in modern times, but speculated that it "might be related to the prevalence of carcinogens in modern societies." David was further quoted as saying, "There is nothing in the natural environment that can cause cancer. So it has to be a man-made disease, down to pollution and changes to our diet and lifestyle."[26]

This finding attracted a great deal of attention, prompting headlines like "Mummies Don't Lie: Cancer Is Modern and Man Made"[27] and "Cancer Caused by Modern Man."[28] It seemed to confirm our worst fears that we had gone down the wrong path by taking up agriculture and settling down to live in cities. The paleo-diet aficionados took comfort in the anticancer nature of their food, while others fretted over the rise in cancer among not only people but their pets. (Meanwhile, the National Cancer Institute notes a slight decrease in new cancer diagnoses, as well as a drop in cancer-caused deaths, over the period between 2003 and 2007, a time when one would not imagine that the pressures or toxins of the civilized world would be decreasing.)[29]

Other scientists, as well as many cancer-related organizations, were not so eager to accept David and Zimmerman's results, or their conclusions that cancer is caused by modern lifestyles. The first problem lies in whether the discovery of few bone-related cancers in the samples reviewed by David and Zimmerman really means that cancer was rare. Many skeletons are incomplete, and of course cancers do not always spread to bone when they originate in soft tissue; the person could die before such metastasis occurs. What's more, most cancers appear in older people, so a sample needs to include enough of the over-fifty crowd to have a hope of detecting many cancers at all. The sample of mummies and skeletons that David and Zimmerman studied had a preponderance of remains from much younger individuals, so not that many cancers should have been expected in the first place.

In a fascinating paper published in 1996, Tony Waldron at University College London asked how much cancer should hypothetically be expected in any given sample of human remains.[30] He then applied his reasoning to a robust ancient data set to see whether his predictions were correct. To generate the predictions, Waldron used the medical literature to create a simple formula for cancer incidence based on the proportion of deaths due to cancer of a particular organ or tissue and the proportion of tumors at that site that metastasize to the bone. This calculation gave him an idea of how much evidence of cancer in bone would be expected out of the total number of cancer cases. He then examined the deaths from cancer in a set of records from 1901 to 1905 and compared these to evidence in burials from the crypt of Christ Church, Spitalfields, used between 1729 and 1857. The period between 1901 and 1905 was selected because records were reliable but tobacco use was not yet prevalent enough to have inflated the cancer rate with associated lung cancer.

Waldron's formula turned out to predict the number of remains that should show signs of cancer with surprising accuracy. Furthermore, Andreas Nerlich and Beatrice Bachmeier from Munich used Waldron's technique on two more ancient samples (an Egyptian population from 1500–500 BC, and a European population from 1400–1800) and found, just as the formula predicted, that only 5 out of the 905 Egyptian skeletons and 13 out of the 2,547 European remains showed signs of cancer.[31] These numbers may sound small, but because only a fraction of the total cancer cases leave evidence in the skeleton, they are exactly in keeping with modern cancer incidence. In other words, cancer is not a curse of modernity, and the reason we may see little evidence of it in ancient remains, or in modern foragers, is simple statistics.

David's remark about no "natural" causes for cancer also sparked criticism from cancer organizations, since solar radiation, radon from rocks, and numerous viruses are known to cause cancer of various sorts. One of the hottest areas in cancer research today is the search for links between cancer and infectious agents. Although it is true that some aspects of modern life—most notably smoking, which causes about a quarter of all cancer cases in the world—are associated with cancer, as Andy Coghlan in the *New Scientist* put it, "Most of them are down to poor lifestyle choices that people can do something about, not, as implied, because they are drowning in a sea of carcinogens from which there is no escape."[32]

Death, perfection, and the evolution of disease

If cancer is old, does that mean we inherited a predisposition to it from our ancestors, along with the ability to breathe air, stand upright, or see colors? And if so, how far back does our vulnerability go? Finally, can information about cancer in other kinds of organisms help us understand not only why species vary in the degree to which they get cancer, but—a much larger question—what determines how long different species live?

Caleb Finch, an eminent biologist who has studied the evolution of aging and senescence in animals ranging from fruit flies to people, suggests that the cancer propensity and longevity might be related.[33] While we have some idea about the life spans of various species, it should come as no surprise that obtaining accurate estimates of cancer rates in other species is at least as hard as getting them from skeletal and other human remains. Animals do get cancer, though the frequency varies (and the notion that sharks do not get cancer is a myth, although the group of fish to which sharks belong does seem to have some unique immune system properties that may help in resisting abnormal cell development). Intriguingly, evidence suggests that chimpanzees, at least, suffer from cancer at a much lower rate than humans, even when they live a relatively long time. Nevertheless, humans, even without modern medical care or hygiene, live far longer than any of our primate relatives.

Finch points out that chimpanzees and the other great apes for which we have information seem to suffer little from the neurodegeneration that accompanies Alzheimer's disease in humans, again even after taking their shorter life spans into account. In other words, a forty-year-old chimp, the equivalent of a human twice that age, is unlikely to suffer from the same age-related diseases that are common in elderly people. Finch further suggests that selection for our long life spans has come at a price. Our immune systems can keep us going for many decades by fending off viruses, bacteria, and other onslaughts, but they also make us prone to inflammation, heart and neurological disease, and cancer. This double-edged sword arises in part because immune system cells regulate many different cell processes, including growth and inflammation, as they distinguish self from nonself and eliminate the latter. But that fine-tuned ability can malfunction, and when it does the result can lead to the unregulated growth that is cancer, as well as a host of other maladies.

This rather literal cost of living, or at least cost of living a long time, has its roots in one of evolutionary biology's most basic questions: Why does any organism grow old and die? Why don't we just keep living, and reproducing, for ever-longer time periods? Remember, natural selection works by favoring those who leave the most copies of their genes, so answers like "we evolved to let others replace ourselves" won't cut it.

The solution to this problem is another variant of the no-free-lunch idea I discussed earlier in the chapter, and in this broader case it is called "negative pleiotropy." Pleiotropy means that a gene will often have multiple effects, acting at different times during the life span and on different organ systems. As a hypothetical example, the same gene that speeds growth might also strengthen bones or weaken arteries. Any genes that enhance reproduction early in life but are detrimental at later ages, once reproduction is over, will still be favored by natural selection. The pleiotropy is negative because of this "higher-now, lower-later" correlation. Such early acting genes are favored because they will be passed along and the individuals bearing them will outreproduce their competitors before the negative effects are felt. By the time the gene that made someone more fertile begins to make its bearer suffer from Alzheimer's (again to give a hypothetical example), the genie is out of the bottle: that person has already left more copies of his or her genes than has someone less prone to brain degeneration but with a smaller reproductive output.

Finch and other biologists believe that selection favored our long life relative to other primates perhaps because it enables us to rear our long-dependent children, but it may also have saddled us with cancer and other age-related ills. Mel Greaves from the Institute of Cancer Research in London refers to our evolutionary heritage as dealing us our position in "the cancer lottery . . . a consequence of the 'design' limitations, compromises and trade-offs that characterize evolutionary processes."[34] In keeping with this perspective, evolutionary biologist Bernie Crespi speculates that any genes aiding survival and growth in childhood will be strongly favored by natural selection, even if they increase diseases later.[35] To those who think that this early survival is not really such a big deal, Crespi reminds us that infections and malnutrition used to kill large numbers of children, and they still can in parts of the developing world. In keeping with his idea, a gene that is associated with higher birth weight, a characteristic that can protect infants from such early threats, also confers susceptibility to a number of autoimmune and other disorders. As Crespi puts it, "The evolution of human disease risk becomes inextricably linked with the evolution of modern humans."[36]

The truth is that diseases have indeed always been with us, modern only in the sense that some of them accompanied our evolution into human beings, not our recent adoption of agriculture or urban living. We all wish we could be healthier, and it is easy to fantasize that before Big Macs, or roads, or houses, we were. But evolution doesn't work that way, with the accomplishment of perfect health or perfect adaptation after some arbitrary period of time. Instead, diseases perfectly demonstrate that life is an endless series of checks and balances, with no guarantees of a happy ending. "What kind of life is that destined to be sick by our own genes?" The answer seems to be, the only kind of life we have.

10

Are We Still Evolving?

A TALE OF GENES, ALTITUDE, AND EARWAX

In 2009, Charles Darwin would have been 200 years old, and his landmark book *On the Origin of Species* reached the 150th anniversary of its publication. To commemorate these events, conferences were organized, panels were convened, and volumes were published. The round numbers of the occasions seemed to generate reflective commentary along the lines of those where-are-they-now pieces about movie stars whose careers peaked decades ago, checking to see whether Darwin's ideas still hold, much the way we periodically try to determine whether Robert De Niro is still a relevant force in film.

Much of the discussion, at least in the popular media, centered on humans, and the topic most frequently addressed—and the question that biologist Jerry Coyne, author of *Why Evolution Is True*, says he is invariably asked when he speaks to the public—was whether humans are still evolving. Some people have strong feelings about the matter; on science writer Carl Zimmer's blog *The Loom*, a commenter thought further human evolution was inevitable but rued the fact: "Why cant [*sic*] our genes know that we dont [*sic*] want to change anymore. We're very content with what they've achieved till now . . . Why does it [*sic*] evolution have to spoil this temporary bliss for such a wonderful intelligent species?"[1] On Cavemanforum.com, at least one reader was less optimistic: "I think people are rapidly changing, but let's face it. This is not a good thing."[2] Another pettishly remarked, "So we may go to all the trouble of evolving and still not even be able to utilize grains in their raw state, that is unfortunate."[3] On a more yearning note, a commenter on the blog *Sports Abode* suggests, "I think we are done evolving but I think the next big change will be wings. We could use some wings."[4]

Running through these remarks is the feeling that evolution is something extra that has been done to us humans, either spoiling or improving on a product that would have existed in some form regardless of its actions. This is erroneous, of course; people are a product of evolution, and so are all other organisms. Yet no one seems to wonder whether sloths decry their inability to walk on the ground, or if alligators could someday live vegetarian lives.

Part of the curiosity about continued human evolution probably stems from our continuing need to see ourselves as different from other animals—a form of hominin exceptionalism, as it were. It isn't necessarily related to one's stance on the origin of humanity; even those who accept that the human species has evolved may believe we are now past all that, having finished up with evolution some time ago, as if checking it off on a grandiose "to do" list. The world of iPads and organ transplants seems far removed from the "nature, red in tooth and claw"[5] that set the stage for Darwin's ideas.

It turns out that although renewed interest in whether modern-day humans are evolving may have been triggered by the *Origin of Species* bicentennial, the question itself arose almost immediately after the book was published. Lawson Tait, a gynecologist and surgeon living in Birmingham, England, was a contemporary of Charles Darwin and enthusiastic promoter of Darwin's ideas, having read *On the Origin of Species* as a student soon after its 1859 publication. Just a decade later, Tait published a paper in the *Dublin Quarterly Journal of Medical Science* with the worried title "Has the Law of Natural Selection by Survival of the Fittest Failed in the Case of Man?"

Tait was commenting on another article, by William Greg, that had appeared in *Fraser's Magazine* with a similarly fretful tone: "On the Failure of 'Natural Selection' in the Case of Man." Both authors feared that the increasing ability of the medical profession to prevent deaths from disease would compromise the culling action of selection, so that, in Greg's words, "those whom [medical science] saves from dying prematurely, it preserves to propagate dismal and imperfect lives." Thus, "if we eradicated disease, the chances of life and death being equal, beings of an equal calibre, in every respect, would be produced."[6] Tait went on to wonder whether medical practitioners should simply give up, allow nature to take its course, and permit only the most vigorous members of society to survive. But he consoled himself with the thought that "no sooner has science overcome one epidemic than another, far more formidable, makes its appearance."[7] Selection therefore continues, but with different criteria as each epidemic arises. Tait concluded that those in his profession should persist, but with the recognition that their labors would often be Sisyphean.

I have to confess that I found Tait's even limited optimism about the ability of late-nineteenth-century physicians to eliminate disease a bit ill founded, given the inadequacy of medicine at the time; most drugs were ineffective and sometimes downright dangerous, and bloodletting was still being performed. Be that as it may, his point about modern technology and its ability to buffer us from the forces of nature that promote selection in other species was echoed by Darwin himself, who cited Tait's and Greg's work in his 1871 book *The Descent of Man and Selection in Relation to Sex*. "There is reason to believe," Darwin mused, "that vaccination has preserved thousands, who from a weak constitution would formerly have succumbed to small-pox. Thus the weak members of civilised societies propagate their kind."[8]

Along with Tait, Darwin nevertheless believed that humans could evolve, and the two men went on to engage in an extended communication, following up a mutual interest in insectivorous plants such as sundews. Tait also became one of Darwin's most ardent fans and supporters. Although the two scientists did eventually meet, the way Tait's biographer J. A. Shepherd describes his subject's behavior, it bordered on stalker-like, or what passed for such in the 1870s, with increasingly frequent letters and invitations that Darwin often answered only briefly or declined outright.[9] Along with Thomas Huxley, however, Tait helped Darwin's ideas become established among the intelligentsia of Victorian England, so his role in the early history of evolutionary theory remains significant.

Violating the law of the jungle?

In a more modern take on the idea of evolution stopping, geneticist Steve Jones from University College London famously suggested several times over the last decade that it has indeed come to a halt, at least in the Western world, and that we should "look around—this is it. Things have simply stopped getting better, or worse, for our species."[10] The argument is similar to Tait's: most children are surviving, even if they have traits such as myopia or susceptibility to measles that would have been disadvantageous just a few hundred years ago. Contraception allows people to determine the number of children they have, further interfering with the different ability to perpetuate one's genes that is inherent in evolution by natural selection. In addition, Jones notes that the genetic mutations that provide fodder for changes in the gene pool are more common in older fathers, because errors in DNA accumulate over time, but younger men are having more children now than they were in the past. Finally, he argues that the large, mobile populations of today mean that everyone's genes are mixing, in effect homogenizing the people of the world into one large group. This means that small groups of individuals can't respond as quickly to localized forces of natural selection. Furthermore, like Tait and his contemporaries, Jones feels that our genes are now freed from the tyranny of environmental pressures such as predators and devastating diseases, simply because we have removed them through technology.

Although it is initially intuitively appealing, many evolutionary biologists—I among them—disagree with this conclusion. The most obvious objection is that while most of the scientists engaged in the discussion live in developed countries, many of the world's people do not. They may not be as much at risk of being eaten by a tiger as their, and our, ancestors were, but children in many places still die of malnutrition and a host of diseases, exposing them to natural selection for the ability to survive such assaults. In addition, people have migrated to new and challenging environments within the very recent past, with concomitant evolution; later in the chapter I discuss one such case: Tibetans adapting to life at high altitude. There is no reason to think that such movements might not result in similar adaptations in the future.

Even in more industrialized societies, however, disease remains a very real threat. The extraordinary medical advances of the past century notwithstanding, epidemics continue to take their toll, with SARS, H1N1 influenza, West Nile virus, and of course HIV as recent examples. As we saw in Chapter 9, the latter already may favor bearers of the *CCR5*-Δ gene variant to become more common in the future. Tait was correct: because pathogens themselves continually evolve new ways to get around the defenses that we, their hosts, put up, the battle with infectious diseases will never be over. The paleo-lifestyle proponents often focus on diet as a central part of human evolution, but disease has arguably left a much larger footprint on our genes.

Medicine and agriculture are both part of the larger human attribute we call culture, along with our ability to create houses, tools, and clothes. These skills and objects allowed humans to occupy parts of the globe that are hostile to most other forms of life, such as the Arctic, and they have changed the playing field for the game of evolution. Jay Stock of the Leverhulme Centre for Human Evolutionary Studies at the University of Cambridge says, "If our culture effectively removes us from environmental stress, then natural selection will no longer occur."[11]

So, does culture insulate us from the environment? Partly, but by no means completely—and it may even make some of those environmental stresses, like crowd-dependent diseases, more apparent. If technology allows us to live in larger societies, the chance of spreading diseases such as tuberculosis or measles, which require a large pool of susceptible victims, increases. And those diseases are very effective agents of selection. In addition, culture itself shouldn't be discounted as somehow different from the rest of our surroundings. As Meredith Small points out, "Culture may not seem a 'natural' force, but because it is part of our environment it is just as natural as disease, weather or food resources . . . humans haven't really changed the rules of natural selection."[12] What's more, culture can itself act as an agent of selection; evolution by natural selection requires organisms to have different numbers of surviving offspring, and few forces are as powerful in influencing our family size as the society in which those people live.

John Hawks, the University of Wisconsin anthropologist mentioned in earlier chapters, took on Jones's contention that younger fathers lack the mutations that will drive evolution. He notes that it is true that, as Jones argues, older men dominated the reproductive pool in the past, partly because older men were better able to afford to marry and reproduce.[13] Older men are more likely to bear alterations in their DNA that can be passed on to their children, and such alterations are the source of new genetic variation.

This disparity in age of who reproduces is not so common today. But a much more salient change is the enormous growth in the human population. More people means more opportunity for mutations—a point also elaborated on by Gregory Cochran and Henry Harpending in *The 10,000 Year Explosion*. As Hawks puts it, "There are a smaller proportion of older fathers now than in 1700, but the absolute number of older fathers is much, much greater."[14] This larger population means a richer source of those novel gene variants, and hence greater scope for evolution.

Drifting along

To fully answer the question of whether we are still evolving, it is important to distinguish between two concepts that are sometimes—incorrectly—used interchangeably: evolution and natural selection. At its core, evolution simply means a change in the frequency of a particular gene or genes in a population, so that, say, in a group of ten hamsters, a gene that makes its bearers wiggle their whiskers starts out in just three individuals, but after several years it is present in six members of the group. The population has evolved, because the proportion of individuals with the whisker-wiggling gene has gone from 30 percent to 60 percent.

The question is, *why* did whisker-wiggling become more prevalent? Here is where the distinction between natural selection and evolution comes in. Biologists commonly recognize four mechanisms by which evolution can occur: genetic drift, gene flow, mutation, and natural selection. Genetic drift is the alteration of gene frequencies through chance events. Imagine that the initial population of hamsters goes for a walk underneath a cliff. Tragedy strikes when a large boulder falls from above on five of the hamsters, killing them instantly. Simply by chance, none of the five unlucky victims carried the whisker-wiggling gene. Once the remaining hamsters reproduce, the population now has a preponderance of whisker wigglers.

When I use a similar analogy in teaching, my students occasionally answer an exam question asking them to define genetic drift with "when a rock falls on some members of a population," but hopefully the point is clear here: being able to wiggle one's whiskers had absolutely nothing to do with the likelihood of being killed. Chance events alone caused the trait to increase in frequency nonetheless. Among humans, genetic drift is thought to have caused some of the changes in face shape over the last few million years, from that heavy-browed look of, well, cavemen, to the more delicate skulls of today. Drift is particularly likely to act in small populations, because random events (like that rock falling) can influence a larger proportion of the group and hence more drastically change the gene frequencies.

Gene flow is simply the movement of individuals and their genes from place to place, an activity that can itself alter gene frequencies and hence drive evolution. Instead of a rock falling on part of our hapless hamster troop, some additional whisker-wiggling individuals could have wandered into the population from another place. To use an example slightly more relevant to the matter at hand, Vikings may have spread the *CCR5-Δ* gene as they moved south through Europe—a factor that could have increased the frequency of that variant apart from any influence of selection. Although Jones suggested that the increased gene flow in the modern world would slow or even halt evolution in the cosmopolitan world of today,[15] Hawks notes that, as with larger populations meaning that more fathers contribute their mutations, more mutations will also be entering the more mobile populations, and once again such increased variability facilitates evolutionary change.[16]

The final way that evolution sans natural selection can occur is via those mutations, changes in genes that are the result of environmental or internal hiccups that are then passed on to offspring.

Such genetic alterations, such as the one that causes Tay-Sachs or other genetic diseases, are usually harmful, simply because random changes to complex machinery are rarely an improvement, which means that their bearer does not get an opportunity to reproduce. A small minority, however, are not harmful, and they, too, reform the genetic landscape and cause a change in gene frequencies.

The suspicion that humans might not be evolving may stem from the misconception, discussed earlier, that evolution is progressing toward a goal. A related fallacy is that all of life has been aimed at the production of humans, the pinnacle of evolution. If humans are seen as an end point, then presumably there is no need for further modification, making continued human evolution a sort of gilding of the anthropological lily. Such a notion is scientifically indefensible, of course; nature has not singled out humans for special treatment, and human beings are not the most recently evolved species on the planet in any event. That last honor, if it can be viewed as such, would likely belong to a virus, a bacterium, or another microorganism, since their short generation times allow them to evolve, almost literally, in the blink of an eye. Humans are no more of an end point to evolution than they are its most recent product.

The bottom line is that evolution in modern-day humans is easily demonstrated according to the criteria of change in gene frequency in our population; numerous surveys of the human genome and comparisons to ancient DNA or the genes of our closest relatives, the great apes, illustrate that the frequencies of many genes have changed. In addition, it simply stands to reason that our genes, like those of other species, are buffeted about by random forces, so we should not expect humans to be special in that regard. But to many people, this answer, that we are evolving because our genes have changed, often through random forces, is unsatisfying. What they really want to know is whether natural selection, the remaining evolutionary mechanism, is still in force. Here, too, the answer is yes.

Plumper, but with lower blood pressure

Upon reflection, it should come as no surprise to hear that the evidence of recent selection is all around us. It may help to do as Mary Pavelka, a primatologist at the University of Calgary in Canada, suggested in a 2005 article: "The question 'Are humans still evolving' should be rephrased as 'Do all people have the same number of children?' "[17] The answer is obviously no, which means that differential reproduction is occurring. If people have genetic differences in their fertility, natural selection has an opportunity to act. The use of contraception complicates matters somewhat, but that opportunity does not go away.

I have already discussed one of the best-documented examples of recent selection in humans: the rapid increase in the ability to digest lactose. But another illustration of our continual evolution is much closer at hand, both in time and space. What's more, uncovering the evidence relied on rather unglamorous techniques that have been around far longer than the flashy genomic technology of the twenty-first century.

The Framingham Heart Study is an ongoing survey of about 14,000 residents of Framingham, Massachusetts, that began in 1948. Sponsored by the National Heart, Lung, and Blood Institute and Boston University, the study is the longest-running multigenerational study of its type. It was originally intended to uncover the risk factors for cardiovascular disease by tracing the lives, and deaths, of multiple generations of Framingham residents. People enrolled in the study are given medical exams every two to four years, and information about their blood pressure, cholesterol levels, and other health markers, as well as basic statistics on height, weight, and the number of children they have, is carefully recorded. Over the last half century, researchers using these data have made a number of important discoveries, including genes associated with

Alzheimer's disease and links between sleep apnea and the likelihood of having a stroke.

The scientists who began the Framingham study were not particularly interested in evolution, but the information from the study turns out to be exactly what biologists who look for evidence of natural selection in any population, human or not, need. The first and second generations of women from the study are now postmenopausal, which means they have already had all their children. A third generation is in the midst of data collection and still reproducing. The information being generated by the Framingham residents is not terribly different from that collected by scientists examining evolutionary change in fruit flies or mice: researchers look for a correlation between the number of offspring an individual has had and her own characteristics, such as body size (the height and weight measures taken for the Framingham study) or health. If women with a certain trait reproduce more, and the trait itself is one that can be passed from parents to offspring, natural selection occurs.

Much as the Grants and their colleagues did with finches on the Galápagos Islands (see Chapter 3), a team of scientists led by Stephen Stearns of Yale University calculated which characteristics were associated with having more children.[18] Using humans instead of birds has its advantages and disadvantages in this type of study. On the plus side, the same individuals could be unquestionably identified and measured over and over again—a difficult task when attempting to catch finches in a net. On the minus side, humans smoke, use medications, and are educated about their health—all components of human existence that are conspicuously absent in avian society and are potential sources of interference in drawing conclusions. What if, for example, smokers happened to have more children because they were wealthier, or highly educated people were more likely to be taller? Luckily, the latter complications could be dealt with in the statistical analysis of the data. Similar accounting methods were used to ensure that early deaths, before people could reproduce, did not bias the sample.

After a complex set of calculations, Stearns and his colleagues found evidence of natural selection on the women in the study (focusing on women is sensible because each has an unambiguous number of children, which is not necessarily true for men). The women of Framingham seemed to be getting shorter and slightly plumper, and both their cholesterol levels and blood pressure were decreasing. The age at which they had their first child was also going down, but the age at which menopause occurred was going up, leading to a longer reproductive period in the total life span. We can only speculate about the causes of these changes, other than to state the obvious: that women with the new set of characteristics tend to have more surviving children than do those with the older set. Certainly, lowered cholesterol and blood pressure are associated with better overall health, but the height and weight alterations are more difficult to explain.

The scientists were then able to use their results to predict the future. Or, put a bit more modestly, they could calculate what the women of Framingham would look like in ten generations, assuming the same associations between fertility and the various health measures. Average weight, for example, is expected to go up by 1.8 kilos (4 pounds), while total serum cholesterol should decrease by 2.35 milligrams per 100 milliliters, or about 3.6 percent. The other traits measured had similarly small changes, leading the authors to conclude, "Natural selection is acting to cause slow, gradual evolutionary change."[19] Of course, as the authors recognize, there is no guarantee that the environment will remain the same over those ten generations, and any environmental changes would throw off the predicted trajectory.

The authors of the study also identified a number of other sources of data that could be used to examine natural selection on human populations. In a review paper published in 2010, the same team noted that while the Framingham Heart Study was one of the earliest initiated, similar long-term surveys are being made of people in Denmark, Gambia, Great Britain, and Finland, as well as several other US states.[20] The surveys have a variety of goals, ranging from collecting data on general health to understanding cardiovascular disease in African Americans, but all are a gold mine of information for evolutionary biologists, as long as they also include data on how many children the participants produce.

A few characteristics were consistent across several of the studies. People, both men and women, who began having children at an earlier age were favored by selection, meaning that their genes were more likely to appear in succeeding generations. In at least one preindustrial population (Finland during the seventeenth through nineteenth centuries) and two postindustrial populations (twentieth-century Australia and the United States), selection also favored women either having their last child at a later age or being older at menopause.

In keeping with these results, a 2011 paper by Canadian researchers found a decrease in the age at which women living on Île aux Coudres, a small island north of Quebec City along the St. Lawrence River, had their first children.[21] The researchers, led by Emmanuel Milot from the University of Quebec in Montreal, reviewed church records for women married after 1799 and before 1940. The population was isolated from the rest of the province and hence provided a nicely demarcated sample. Over the 140 years surveyed, the age at first reproduction went from twenty-six to twenty-two. This change had a genetic component, probably because characteristics related to child-bearing ability, such as the growth rate or age at first menstruation, are themselves heritable. In addition, the average number of children per woman went from three to four during the same period.

All of these findings suggest that what Stearns and his colleagues call "the temporal window of reproductive opportunity"[22] is growing larger, enabling people to have children throughout more of their lifetime. Of course, people may not take advantage of that open window, which means that culture, again, may end up ameliorating the effects of selection in the long term.

The results of the review by Milot and colleagues also put to rest the notion that modern medicine and public health improvements such as sewage treatment have halted evolution in humans. Recall Pavelka's question: Are all people having the same number of children? Whether or not survival increases, differences in reproduction among individuals are what matter. Culture, in the form of contraception or assisted reproduction, may complicate the story, but ultimately culture just forms part of the environmental backdrop against which evolution occurs. Environments always vary in the intensity of selection: if a storm kills off half the population, as it did with Bumpus's sparrows, selection is extreme, while if a shift in seed size means that some birds are able to lay one egg fewer than usual, selection is weaker. Culture may change the strength of natural selection, or the traits upon which its action is most apparent, but it will not completely eliminate it. This means that instead of dismissing evolution in humans as too subject to the vagaries of culture to consider, new research areas present themselves. For example, how does *in vitro* fertilization with donor sperm or eggs change the rate of evolution? What about the use of surrogates? These questions are part of a new frontier in biology.

Apropos of this point, although natural selection is acting in modern human populations, it's important to recognize that it does not act with the same intensity, or even in the same direction, everywhere. Studies measuring selection on height offer an example: In some places, such as Gambia, women's height was increasing; in others, such as Poland or some parts of the United States, it was decreasing. In yet others, selection was stabilizing, meaning that the average height was not projected to change much at all.[23] This set of results is typical for a large, heterogeneous population of organisms. Some environments favor being tall, some favor being short, and the same environment can do both at different times. The review authors suggest that women in preindustrial populations are under selection to become taller, while those in more industrialized places, such as Massachusetts, have more children if they are shorter.[24] We do not know the reason for this difference; perhaps taller women do better at the physically demanding tasks of a less technologically dependent society. Regardless, environmental variability drives the difference in selection pressure. More generally, this variation underscores the point that there is no single optimum state for humanity—being tall can be better than being short, or vice versa, for different individuals under different circumstances.

Evolving up where the air is clear—and thin

Elegant though Stearns' study is, it did not examine genetic change directly; that was not its intent. None of the researchers examined DNA itself, instead relying on the more indirect, if time-honored, methods of studying the changes in the characteristics of a population to infer genetic change. We already know that height and blood pressure are partly inherited from our parents, so inferring that changes in averages of those traits is due to selection is not too far a leap. But other recent research has taken another step, revealing the exact nature of human evolution in action by examining the genes themselves.

A 2010 *New York Times* article by Nicholas Wade was headlined "Scientists Cite Fastest Case of Human Evolution." Everybody loves the winner of a race, and here the winners were the high-altitude natives of Tibet, where life at 13,000 feet above sea level does not produce the usual symptoms of mountain sickness, a malady arising from the body's unsuccessful attempt to counteract the lack of oxygen in the air. In non–mountain dwellers, the body compensates for low oxygen levels by increasing its production of red blood cells. Red blood cells are in turn the carriers of hemoglobin, the molecule that binds to oxygen and carries it to the organs and tissues. The increase in red blood cells thickens the blood and causes a number of health problems, ranging from headaches and insomnia to difficulty breathing and brain swelling, and in the long term it can reduce fertility and cause women to have smaller babies.

The Tibetans, however, seem to have overcome these difficulties. Unlike other people, they do not have an elevated amount of hemoglobin in their blood, but instead have higher breathing rates at rest without experiencing any ill effects. In contrast, when Han Chinese, the majority ethnic group in China, live at the same high altitude as the Tibetans, they exhibit signs of the kind of chronic mountain sickness that most Westerners would experience under similar conditions. How do the Tibetans cope, and why the difference between the two groups of people?

Human beings evolved more or less at sea level, and the extreme environments of high mountains were colonized much more recently, with current estimates ranging from 11,000 years ago for the Andes and perhaps 3,000–6,000 years ago for the Tibetan Plateau. Interestingly, Andean natives do not show the higher respiratory rate seen in Tibetans, and they also have high hemoglobin concentrations in their blood, in contrast to the mountain people of Tibet. Although both groups of people thrive at high altitudes, they seem to have different solutions to similar problems.

The genetics of the Andean natives have not been well studied, but independent groups of researchers recently demonstrated that Tibetans exhibit unique genetic adaptations to living at high elevations. In the first example, a group of scientists from China and the United States scanned fifty unrelated individuals from Tibet and compared their genomes with those of forty people of ethnic Han heritage.[25] Using a different approach, a team led by Tatum Simonson of the University of Utah and RiLi Ge of Qinghai University in China looked for genes in Tibetans, Chinese, and Japanese that seemed to have been subject to selection for the ability to deal with low oxygen levels.[26] Variants of a gene catchily named hypoxia-inducible factor 2-alpha, or *HIF2-α* for short, as well as two genes that modify its effects, were associated with a lower concentration of hemoglobin in the blood of the Tibetan highlanders. Having lower levels of hemoglobin means that the Tibetans avoid the problems that come along with more red blood cells.

If avoiding the overproduction of red blood cells is more advantageous, why doesn't the human body do that naturally, rather than vainly ramping up hemoglobin levels, which creates more problems than it solves? The answer illustrates the tinkering nature of evolution that I mentioned in Chapter 6. According to Jay Storz from the University of Nebraska, increasing one's hemoglobin concentration in response to decreased oxygen levels may have evolved to counteract anemia, which also causes a decrease in the amount of oxygen available to the body.[27] But altitude sickness and anemia differ in a crucial characteristic: the latter can be somewhat ameliorated by increasing hemoglobin concentration because it stems from an inability of the blood itself to transport oxygen. With altitude sickness, the blood is less saturated with oxygen because less oxygen is available in the air to begin with, and hence increasing hemoglobin concentration actually prevents oxygen from reaching tissues.

Evolution, of course, doesn't know why the blood levels of oxygen have diminished; in our ancestral sea-level environment, a rule of thumb that says, "When oxygen is low, increase hemoglobin" would work perfectly well. In the Tibetans, however, an individual who exhibited a mutation that did not provoke such a response when oxygen levels were low would actually survive and reproduce better than the more usual variant would. This, of course, is exactly what appears to have happened. In the Andes, the mutation may not have arisen, and the two populations of mountain dwellers could differ simply because different raw material was available for selection to act upon.

The media made much of the "fastest evolution ever" sound bite, in the process highlighting discussions of exactly how long ago the Tibetan Plateau was populated. If the number of years comes in at higher than 7,500, then the prize will go to the evolution of lactase persistence, but if it is the 3,000 or so postulated by the geneticists, then Tibetan altitude adaptation will be the clear winner. Archaeologists, who rely on evidence such as preserved hand- and footprints, remains of tools, or butchered animal bones, tend to favor a longer settlement history, but the matter is still not resolved. Among geneticists, a few thousand years more or less is well within the bounds of their estimation, so they are not much bothered either way.

Earwax, and sweeping through time

Regardless of who wins the fastest-evolution race, the number of demonstrations of recent evolution and selection in the human genome is mounting: lactase persistence, amylase starch digestion, malaria resistance, adaptation to high altitude. We could keep adding to the list, and we are (wait till I get to the earwax), but now that the human genome has been sequenced and it is possible to analyze large sections of DNA at a time, many scientists are taking a different approach. By examining sequences that appear to have been inherited as a block, it is possible to detect the genes that have been subject to recent selection.

What's more, scientists can often distinguish between so-called positive and negative selection. While natural selection results in individuals predominating if their genes are more suited to the environment, that process can happen in either of two ways. With negative selection, new harmful mutations are culled from the genome, changing the gene frequencies in the population (hence causing evolution), but doing so by subtraction, not addition. Negative selection is the reason that many of the genetic regions associated with manufacturing proteins are so similar among species: making proteins is fiddly work, and small deviations can be disastrous, so any mutations that altered the ancestral plan would likely not survive. Negative selection is the Grim Reaper of evolution, relentlessly removing, according to recent studies, up to three-quarters of the mutations that arise, though most of that elimination occurs very early in life, even before a fertilized egg implants in the uterus.

Positive selection seems more creative and somehow cheerful, acting to increase the frequency of a new or previously rare mutation that renders its bearer more likely to survive and reproduce. As a 2007 paper by geneticists from Cornell University and the University of Copenhagen notes, positive selection has received a great deal of interest from scientists "because it provides the footprints of evolutionary adaptation at the molecular level."[28]

We can detect selection by looking for so-called selective sweeps, areas in the genome where chunks of DNA appear to have been inherited *en masse*. I discussed the basic idea in Chapter 5, where I pointed out that it is possible to determine whether genes associated with digesting different foods are found in people whose heritage came from various parts of the world. When selection favors a certain gene, its neighbors are carried along with it, and the resulting block of DNA becomes more homogeneous than expected because the nearby genes are hitchhiking along. By determining the likely function of genes caught up in such selective sweeps, it is possible to see which genes are subject to more rapid evolution. In the figure shown here, each person has a different set of genes, which would ordinarily be recombined in succeeding generations.

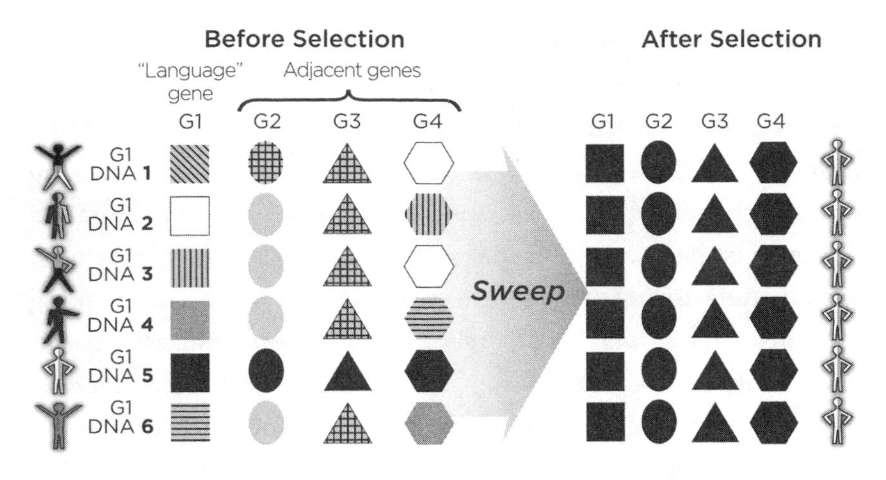

The aftermath of a selective sweep. All of the people have different gene combinations that they would ordinarily pass on to their descendants. But if variant 5 is favored, it will rapidly appear in a disproportionate number of people in the next generations, accompanied by the genes that are close to it, even if those genes are not themselves advantageous. (Adapted from a figure by Rebecca Cann)

But because the language gene 5 is advantageous, the surrounding genes will also all be the same following selection.

Searches for selective sweeps have yielded a number of areas in the genome where selection seems to have acted relatively recently, meaning within the last few tens of thousands to hundreds of thousands of years. Many of those rapidly evolving genes are associated with disease resistance, as I noted in Chapter 9. The "arms race" that occurs as a pathogen's genes evolve new ways to get around the defenses of the host, and the host counters with ever-more effective defenses, provides a good proving ground for advantageous gene variants.

A recent survey of African American, European American, and Chinese gene samples by many of the same researchers who wrote the review mentioned earlier found evidence of selective sweeps at 100 regions in the genome that were located near a gene with a known function.[29] Skin pigmentation, odor detection, nervous system development, and immune system genes appeared to show the most signs of recent evolution. In addition, the scientists found that the Chinese and European American populations showed more selective sweeps than did the African American population. This result is expected if the humans who left Africa experienced new selective pressures as they colonized environments with climates, foods, and diseases that were radically different from those of their homeland. The new environments provided strong selection and hence sweeps should be easier to detect. In all, the researchers concluded that as much as 10 percent of the human genome shows the effects of such selective sweeps.

Although enthusiasm for the search for selective sweeps has been high for the last several years, a few scientists have sounded a note of caution. Jonathan Pritchard and Anna Di Rienzo titled a 2010 paper "Adaptation—Not by Sweeps Alone" to argue that the old-fashioned way to detect selection, à la the Framingham study approach, may have been overlooked in the frenzy to use shiny new genomic tools. They do not suggest a Luddite approach to genetics, but rather point out that many evolutionary changes occur through the interaction of many genes at different places throughout the genome. Therefore, simply looking for places in the genome where blocks of genes have been inherited together risks missing the more subtle, but no less rapid, changes that have happened over time.

Taking matters further, a group of scientists from Chicago, Israel, and the United Kingdom reanalyzed the DNA sequences of 179 people from four populations, with the aim of determining whether the low genetic diversity surrounding selected genes was really a good indicator of a selective sweep.[30] The rationale behind searching the genome for the areas of DNA that are inherited as a package is, as I described earlier, that the accompanying genes get swept along with the gene that's being selected, yielding areas with less variability than one would expect. But how much variability does one expect in the first place, and could other mechanisms account for patches of low diversity in the genome?

The scientists, led by Ryan Hernandez, now an assistant professor at UC San Francisco, discovered that the low-diversity segments were actually distributed in several places throughout the genome, not just surrounding particular types of genes. The hitchhiking genes, in other words, were not necessarily with the same driver that had picked them up on the highway. In addition, although the geographic variation in gene frequencies was still present in the samples studied by Hernandez and colleagues, so that African samples differed from European samples, the differences were often subtle, and a matter of degree rather than kind. The researchers don't deny the occurrence of selective sweeps, but along with Pritchard and Di Rienzo, they caution against attributing too much importance to this mechanism, and against assuming that such sweeps were the main way in which human evolution occurred.

Usually, even when a gene or stretch of DNA that has recently changed frequency in a population can be identified, the exact function of that gene remains a mystery. It may be possible to say that the gene is part of an immune response, or that it is important in protein synthesis, but more specific examples, aside from the oft-mentioned lactase persistence and a few others, tend to be thin on the ground. A recent exception occurred in 2006, but the assignment of a small piece of DNA to its effect in the human body may leave people less than overwhelmed: a team of Japanese researchers identified the tiny genetic change responsible for the type of earwax a person harbors.[31]

If your earwax knowledge is a bit scanty, you may not know that earwax comes in two distinct forms: wet and dry. The dry form is most common among East Asians, while the wet form, described by the researchers as "brownish and sticky," predominates in Africa and Europe. In southern and central Asia, the two forms occur in more or less equal proportions. The wet form is therefore likely to be the ancestral form, since humans evolved in Africa, with the dry form arising later, probably in northern Asia, given that the northern Chinese and Korean samples exhibited it. Moving south through Asia, the dry form becomes less common, perhaps because people with that form intermarried with southern Asians carrying the wet-earwax gene. Native Americans also exhibit the dry form, consistent with the idea that their ancestors were originally from Asia and came across the Bering Strait via Siberia.

The form of one's earwax does not seem to matter much in the greater scheme of things, so far as we can tell. It does keep dirt and insects out of the ear canal, but no evidence suggests that one form or another is more efficient at this task. The genetic change may have come about through genetic drift, or because it is linked to another, more significant characteristic. The scientists who conducted the study speculate that because earwax type is associated with the amount of sweating, and hence body odor, these traits and not the earwax itself might have been the targets of selection. Asians tend to sweat less than Europeans, and the authors suggest that this attribute might have been advantageous in the cold climate where the ancestors of northeast Eurasians lived. This part of the story is, of course, difficult to substantiate, but the detail with which the change in DNA was documented underscores scientists' increasing ability to trace evolution in our recent past. And the earwax story offers hope that scientists will be able to meticulously track a single change to its adaptive advantage in more cases, enabling us to determine exactly how important big selective sweeps have been in our history.

The rapid river of change

The humble earwax gene and the other examples I have provided in this book are only a tiny fraction of the evidence for recent human genetic change. And "recent" can mean 3,000 years, 10,000, or

200,000, depending on one's perspective. For that matter, even trying to rank one or another trait as "most recently evolved" says more about the human tendency to keep score than it does about evolution itself. As I mentioned earlier in the chapter, no prizes are awarded for "most recently evolved," and if they were, microbes would be the clear winners. But it seems to be a human tendency, evolved or not, to want to be good at things; in a 2009 *Scientific American* article musing on the fate of *Homo sapiens*, University of Washington paleontologist Peter Ward reassures us that "humans are first-class evolvers."[32]

Breathless headlines aside, what is wrong with trying to excel at evolution, or with being a tad competitive about which characteristic evolved most recently? The problem is that this kind of goal-directed thinking still presents evolution as a process with an end point; being good at something implies a skill that can be mastered and, once mastered, allows us to move on to something else. But of course our ancestors were also evolving, and as I discussed in Chapter 1, no organism gets to a point of perfect adaptation, heaves a sigh of genetic relief, and stops. All of our characteristics, and all of our genes, are subject to change at different rates, making the declaration of a finished product impossible.

This does not mean that evolution acts in the same way on all creatures, or on all aspects of them—or that all adaptations are possible. Scientists do not talk about being "good at" evolution, but they do consider what is rather awkwardly termed evolvability. "Evolvability" refers to the capacity of an organism to generate the raw material of evolution—namely, heritable variation—so that natural selection can act. More variation means more opportunities to survive a flood, or find more mates, or change bill size. Some people confine discussion of evolvability to occurrences within populations, like a change in coat color from white to brown in mice living in light or dark soil, while others use it to explain how evolutionary novelties, such as wings, arise in entire groups of organisms. Evolvability is not a virtue, but a property of organisms and their genes, and it may fluctuate over time.

The Greek philosopher Heraclitus was supposed to have said, "No man ever steps in the same river twice." He was known as "Heraclitus the Obscure," in part because many of his works were difficult to interpret. Indeed, some people suggest that this saying means that nothing retains its identity, while others claim Heraclitus meant that because life is constantly changing, one can always reinvent oneself. As applied to evolution, one could interpret his remark to mean that organisms are responding to a continually changing environment, the river, and hence evolution itself is continual too. To truly emulate evolution, however, the river would need to be unending, and the foot that stepped into it would need to be different with every attempt at wading. Not only are our lives different from those of our Pleistocene ancestors, but their lives were different from those of their ancestors as well, and on it goes.

Relinquishing the paleofantasy

From our diets to our dating habits, paleofantasies are everywhere, not only on websites yearning for a more natural lifestyle but in places like the *Chronicle of Higher Education*, which recently ran an article on why people like crispy food that invoked our evolutionary predilection for the crunchy delights of insect snacks (apparently the sound made when eating such foods also led to their having "a privileged place in the brain").[33] And the paleofantasies go beyond simply making up "just-so stories," as the late evolutionary biologist Stephen Jay Gould called them, about the evolutionary significance of everything. We are not only constructing scenarios of why our foibles might have been adaptive; we are fossilizing evolution itself.

A simpler life with more exercise, fewer processed foods, and closer contact with our children may well be good for us. But we shouldn't seek to live that way because we think it emulates our ancestors. We can mimic the life of a preindustrial, or preagricultural, society only in its broadest sense. Rather than trying to use our past to proscribe our present, or our future, we can use it as a way to understand where we came from. Paleofantasies call to mind a time when everything about us—body, mind, and behavior—was in sync with the environment. But as the previous pages have shown, no such time existed. We, and every other living thing, have always lurched along in evolutionary time, with the inevitable trade-offs that are a hallmark of life. We have nothing to lose by giving up our paleofantasies, and everything to gain.

For one thing, we gain some relief from that nagging sense that we are mismatched to our environment, doomed to flounder in a modern world to which we are not suited. True, the specifics of twentieth- or twenty-first-century life are novel: computers, high-rise buildings, cell phones. But instead of denouncing modern living as unsuitable to our Stone Age genes, we need to figure out just what parts of that life send us too far out of our evolutionary zone of tolerance, and to do that we need data, not blanket condemnation. Usually, we can make broad generalizations and do not need to sweat the details. For example, sedentary living is clearly linked to poor health, but we do not have to emulate a mammoth-spearing caveman to remedy the problem. We just need to get up off the couch.

For another thing, we can stop looking for models of our ancient selves in contemporary foraging peoples or the great apes. Everything alive today is just as evolved as every other organism, and nothing mirrors human history in its entirety. We are all related, of course, and we can see how humans, or other primates, respond to different selection pressures in different environments, but no species has a premium on being best adapted to its surroundings.

Relinquishing our paleofantasy also helps us feel more connected to other organisms; just because our lives seem different from those of our paleo ancestors, or our great-ape relatives, does not mean we and they have not been subject to the same forces of evolution. We can even be happier without feeling that we are bucking the system; an article in *Self* magazine on becoming more optimistic quoted neuropsychologist Rick Hanson: "Because the human brain evolved during a time when danger was everywhere, it has a built-in negativity bias. For humans to survive under these dire straits, the brain regions responsible for detecting threats had to be turbocharged. Evolution favored those who were able to react to danger with lightning speed."[34] But this Chicken Little attitude is not uniquely human (and I can attest from personal experience that real chickens do, indeed, behave as though danger lurks in every corner, including corners with nothing more threatening than a food bowl). All animals evolved when danger was everywhere, because, well, danger *is* everywhere. But humans didn't evolve to be any more negative than the next species, and being alert to threats does not make us naturally, or unusually, glum.

Giving up the paleofantasies lets us appreciate that all environments, old and new, leave their mark. As with the baby's foot altered by shoes in modern times and by barefoot walking in ancient times, or the different microbes that flourish in our guts depending on our diets, whatever we choose has consequences, and choices have to be made. We do not have genes plunked wholesale into one environment or another, whether Paleolithic, medieval, or industrial; we have genes that respond to that environment and to each other. A comment on *Mark's Daily Apple*, the blog that features Grok, cautioned, "The major difference we face now compared to our ancient ancestors is that except in rare instances change comes at us more quickly today."[35] Perhaps. But change is continuous, and those ancient ancestors encountered changes as well; they domesticated animals, they grew crops, and they dealt with diseases. Change does not have to mean disaster. Sometimes it just creates more change.

Acknowledgments

I am always grateful to my colleagues and the scientists all over the world who share their data and stories, give me advice, and try to keep me (sometimes successfully and sometimes not) out of trouble. This gratitude is particularly pronounced for this book, as I have ventured into territory usually reserved for anthropologists, and that is ground often marked by acrimony and rancor. The anthropologists whom I have been lucky enough to consult have nonetheless been unfailingly helpful and kind. I am especially grateful to Gene Anderson, Rob Boyd, Becky Cann, Greg Downey, Kristen Hawkes, John Hawks, Rosemary Joyce, Sang-Hee Lee, Joan Silk, and Tim White. Leslie Aiello deserves a special mention for her willingness to let me use the word "paleofantasy," though I have stretched it far past her original intent. Becky Cann heroically saved me at the last minute with a terrific figure of a selective sweep. A number of biologists, including many of my colleagues from the Department of Biology at UC Riverside, have influenced my thinking about rates of evolution. I, along with many of the ideas in this book, found a gracious reception at Uppsala University in Sweden and the School of Animal Biology at the University of Western Australia.

I thank my writer friends, who inspire me in ways large and small: Deborah Blum, Susan Maushart, Virginia Morell, Susan Straight, and Carl Zimmer. Angela von der Lippe, Laura Romain, and Anna Mageras are wonderful editors who care about books, an increasingly rare breed. My agent Wendy Strothman and Lauren MacLeod from the Strothman Agency are always helpful and encouraging, and I appreciate Lauren's adeptness with social media, even if I am still somewhat paleo in my own use of it. Stephanie Hiebert did a heroic job of copyediting. Finally, John Rotenberry has been a stalwart and much appreciated supporter.

Notes

Introduction

1. Author visit to Anders Götherström's laboratory at Uppsala University, Sweden, December 2009.
2. Holland, "Cavewoman's Guide to Good Health."
3. Ehkzu, March 6, 2010 (8:39 p.m.), comment on R. C. Rabin, "Doctors and Patients, Not Talking about Weight," *Well* (blog), *New York Times*, March 16, 2010, http://well.blogs.nytimes.com/2010/03/16/doctors-and-patients-not-talking-about-weight/?apage=4#comment-495227.
4. ACW, August 20, 2008 (12:11 p.m.), comment on T. Parker-Pope, "Why Women Stop Breast-Feeding," *Well* (blog), *New York Times*, August 15, 2008, http://well.blogs.nytimes.com/2008/08/15/why-women-stop-breast-feeding/?apage=8#comment-51267.
5. tman, July 16, 2010 (11:26 a.m.), comment on G. Reynolds, "Phys Ed: The Men Who Stare at Screens," *Well* (blog), *New York Times*, July 14, 2010, http://well.blogs.nytimes.com/2010/07/14/phys-ed-the-men-who-stare-at-screens/?apage=8#comment-548092.
6. Balter, "How Human Intelligence Evolved."
7. Shubin, *Your Inner Fish*.
8. Wade, *Before the Dawn*, 70.
9. Zuk, Rotenberry, and Tinghitella, "Silent Night"; Tinghitella and Zuk, "Asymmetric Mating Preferences."
10. Simpson, *Tempo and Mode in Evolution*, xv.

11. Coyne, *Why Evolution Is True*.
12. McKie, "Is Human Evolution Finally Over?"
13. Cochran and Harpending, *10,000 Year Explosion*.

Chapter 1: Cavemen in Condos

1. Goldstein, "New Age Cavemen and the City."
2. "Diet Fad from the Stone Age."
3. Hawks, "Cavemen Are Happy."
4. Goldstein, "New Age Cavemen and the City."
5. KP, August 22, 2007 (8:53 a.m.), comment on topic "So Close and Yet So Far. . .," *Caveman Forum*, http://cavemanforum.com/research/so-close-and-yet-so-far/msg2590/#msg2590.
6. Simopoulos and Robinson, *Omega Diet*, 24.
7. Jacobsen, "Essential Caveman Lifestyle."
8. Adam, November 4, 2010 (11:51 a.m.), comment on C. McDougall, "Born to Run the Marathon?" *Well* (blog), *New York Times*, November 4, 2010, http://well.blogs.nytimes.com/2010/11/04/born-to-run-the-marathon/?apage=4#comment-594535.
9. Author conversation with Greg Downey, September 2010.
10. One example can be found at http://www.paleojay.com/2011/12/getting-rid-of-eyeglasses.html.
11. Wrangham, *Catching Fire*.
12. Gibbons, "Lucy's Toolkit?"
13. Ibid.
14. Boyd and Silk, *How Humans Evolved* (5th ed., 2009), 279.
15. Milius, "Tapeworms Tell Tales."
16. Green et al., "Draft Sequence of the Neandertal Genome."
17. Boyd and Silk, *How Humans Evolved* (6th ed., 2012), 292.
18. Enard et al., "Molecular Evolution of FOXP2."
19. Edward Hagen, "What Is the EEA and Why Is It Important?" *Evolutionary Psychology FAQ*, last modified September 8, 2004, http://www.anth.ucsb.edu/projects/human/evpsychfaq.html.
20. Rosemary Joyce, e-mail message to author, November 11, 2010.
21. Bower, "Strange Case of the Tasaday."
22. Rosemary Joyce, e-mail message to author, November 11, 2010.

23. *John Hawks Weblog: Paleoanthropology, Genetics and Evolution*, http://johnhawks.net.
24. Ibid., under "humor," http://johnhawks.net/weblog/topics/humor.
25. Hawks, "Neandertal Stories on Parade."
26. Glendinning, "Neanderthals: Not Stupid."
27. Keim, "Neanderthals Not Dumb."
28. "Stone Me."
29. Martin, "We've All Suspected," cited in Hawks, "Ozzy Osbourne."
30. Smith et al., "Dental Evidence"; "Evolution: Neanderthals Matured Fast."
31. Sponheimer and Lee-Thorp, "Isotopic Evidence."
32. Alleyne, "Neanderthals Really Were Sex-Obsessed Thugs."
33. "Neanderthals Had a Naughty Sex Life."
34. Nelson et al., "Digit Ratios Predict Polygyny in Early Apes."
35. A good summary can be found in Manning, *Digit Ratio*.
36. "Cavemen Randier Than People Today."
37. Joyce, "Fingering Neanderthal Sexuality."
38. Nelson et al., "Digit Ratios Predict Polygyny in Early Apes."
39. Kratochvíl and Flegr, "Differences in the 2nd to 4th."
40. Hawks, "'Naughty Neandertals' Did What?"
41. Wrangham and Pilbeam, "African Apes as Time Machines."
42. Wrangham and Peterson, *Demonic Males*.
43. Zihlman, "Paleolithic Glass Ceiling."
44. White et al., "*Ardipithecus ramidus*."
45. Wrangham and Pilbeam, "African Apes as Time Machines."
46. Brown, *Human Universals*, 5.
47. Ibid., 6.
48. Joyce, "Do Our Ancestors Walk among Us?"

Chapter 2: Are We Stuck?

1. Radosavljevic, "Stone Age Sights."
2. Cosmides and Tooby, "Evolutionary Psychology."
3. Cordain, *Paleo Diet*.
4. Diamond, "Worst Mistake."
5. Cochran and Harpending, *10,000 Year Explosion*.
6. Diamond, "Worst Mistake."

7. O'Connell, "Is Farming the Root of All Evil?"
8. Wells, *Pandora's Seed*, 90.
9. Feeney, "Agriculture."
10. General information about the beginning of agriculture comes from "Agriculture and Food," *Food Encyclopedia*, *Huffington Post*, accessed October 2010, http://www.huffingtonpost.com/encyclopedia/definition/agriculture%20and%20food/23.
11. Denham, Iriarte, and Vrydaghs, *Rethinking Agriculture*, 6.
12. Cordain, *Paleo Diet*, 3.
13. Harris, "Agriculture, Cultivation, and Domestication," 30.
14. Wells, *Pandora's Seed*, 24.
15. Lee, *!Kung San*.
16. Diamond, "Worst Mistake."
17. Kaplan et al., "Theory of Human Life History Evolution."
18. Ibid.
19. Diamond, "Worst Mistake."
20. Gage, "Are Modern Environments Really Bad for Us?"
21. Bowles, "Did Warfare?"
22. Singer, "Is Violence History?"
23. Cochran and Harpending, *10,000 Year Explosion*, 65.
24. Hawks, "Why Human Evolution Accelerated."
25. Cochran and Harpending, *10,000 Year Explosion*.
26. Powell, Shennan, and Thomas, "Late Pleistocene Demography."
27. Wells, *Pandora's Seed*, 53.
28. Cosmides and Tooby, "Evolutionary Psychology."
29. Ibid.
30. Edward Hagen, "What Is the EEA and Why Is It Important?" *Evolutionary Psychology FAQ*, last modified September 8, 2004, http://www.anth.ucsb.edu/projects/human/evpsychfaq.html.
31. Strassmann and Dunbar, "Human Evolution and Disease."
32. Britten, "Divergence between Samples."
33. Marks, *What It Means to Be 98% Chimpanzee*, 29.
34. Cordain, *Paleo Diet*, 9.
35. Rebecca Cann, e-mail message to author, December 2010.
36. Carroll, "Genetics and the Making of *Homo sapiens*."
37. Ibid.
38. Tooby and Cosmides, "Past Explains the Present."

39. Irons, "Adaptively Relevant Environments."
40. Ibid.
41. Ibid.
42. Wright, "Roles of Mutation."

Chapter 3: Crickets, Sparrows, and Darwins—or, Evolution before Our Eyes

1. Hendry and Kinnison, "Perspective."
2. Bumpus, "Elimination of the Unfit."
3. Haldane, "Suggestions as to Quantitative Measurement."
4. Gingerich, "Rates of Evolution"; Gingerich, "Quantification and Comparison."
5. Robinson Jeffers, "The Beaks of Eagles," PoemHunter.com, submitted April 12, 2010, http://www.poemhunter.com/poem/the-beaks-of-eagles.
6. Grant and Grant, "Unpredictable Evolution"; Seger, "El Niño and Darwin's Finches."
7. Ogden Nash, "The Guppy," PoemHunter.com, submitted January 13, 2003, http://www.poemhunter.com/poem/the-guppy.
8. Reznick et al., "Evaluation of the Rate of Evolution."
9. Reznick, Ghalambor, and Crooks, "Experimental Studies of Evolution in Guppies."
10. Reznick, Ghalambor, and Crooks, "Experimental Studies of Evolution in Guppies"; Reznick et al., "Evaluation of the Rate of Evolution."
11. Ibid.
12. Berthold et al., "Rapid Microevolution."
13. Barrett et al., "Rapid Evolution of Cold Tolerance."
14. Carroll et al., "And the Beak Shall Inherit"; Carroll, "Facing Change."
15. Schoener, "Newest Synthesis."
16. Hendry and Kinnison, "Perspective."
17. The Ten Thousand Toads Project. *Turtle Care Sunshine Coast*, accessed July 16, 2012, http://www.turtlecare.com.au/10k-toads-project.php.
18. "Kill Cane Toads Humanely: RSPCA," *Animals Australia*, February 19, 2011, http://www.animalsaustralia.org/media/in_the_news.php?article=1948.
19. The Kimberley Toadbusters, http://www.canetoads.com.au.
20. Phillips and Shine, "Adapting to an Invasive Species." See also http://www.canetoadsinoz.com.

21. Phillips, Brown, and Shine, "Evolutionarily Accelerated Invasions."
22. "Cane Toad Evolution," CaneToadsinOz.com, accessed July 16, 2012, http://www.canetoadsinoz.com/cane_toad_evolution.html.
23. Ibid.
24. Coltman et al., "Undesirable Evolutionary Consequences"; Allendorf and Hard, "Human-Induced Evolution."
25. Eldridge, Hard, and Naish, "Simulating Fishery-Induced Evolution."
26. Wolak et al., "Contemporary, Sex-Limited Change."
27. Rudolf, "Speedy Evolution, Indeed."
28. Wirgin et al., "Mechanistic Basis of Resistance to PCBs."
29. Ibid.
30. Elmer et al., "Rapid Sympatric Ecological Differentiation."
31. Halfwerk et al., "Negative Impact of Traffic Noise."

Chapter 4: The Perfect Paleofantasy Diet: Milk

1. Robert M. Kradjian, "The Milk Letter: A Message to My Patients," Notmilk.com, accessed February 2011, http://www.notmilk.com/kradjian.html.
2. Robert Cohen, "Detox from Milk: Seven Days," Notmilk.com, accessed February 2011, http://www.notmilk.com/detox.txt.
3. interfool, January 1, 2011 (6:59 p.m.), comment on T. Bilanow, "Salads with Crunch, Sweetness and Zest," *Well* (blog), *New York Times*, December 31, 2010, http://well.blogs.nytimes.com/2010/12/31/salads-with-crunch-sweetness-and-zest/#comment-616849.
4. Warren Dew, February 1, 2011 (7:36 p.m.), reply #3 on topic "Whats [*sic*] Wrong with Cheese?" *Caveman Forum*, http://cavemanforum.com/diet-and-nutrition/whats-wrong-with-cheese/msg47077/#msg47077.
5. Schiebinger, "Why Mammals Are Called Mammals."
6. Information about milk types comes from the University of Guelph's Department of Food Science, http://www.uoguelph.ca/foodscience/content/table-3-composition-milk-different-mammalian-species-100-g-fresh-milk.
7. Bloom and Sherman, "Dairying Barriers."
8. Beja-Pereira et al., "Gene-Culture Coevolution."
9. Boyd and Silk, *How Humans Evolved*.
10. Gerbault et al., "Evolution of Lactase Persistence."
11. Alan R. Rogers, notes for lecture titled "Evolution of Lactase Persistence,"

November 16, 2009, http://content.csbs.utah.edu/~rogers/ant5221/lecture/lactase-2x3.pdf.

12. Gerbault et al., "Impact of Selection and Demography."
13. Anderson and Vullo, "Did Malaria Select?"
14. Tishkoff et al., "Convergent Adaptation."
15. Ingram et al., "Lactose Digestion."

Chapter 5: The Perfect Paleofantasy Diet: Meat, Grains, and Cooking

1. Revedin et al., "Thirty Thousand-Year-Old Evidence."
2. Henry, Brooks, and Piperno, "Microfossils in Calculus."
3. Henry et al., "Diet of *Australopithecus sediba*."
4. Hirst, "Grinding Flour."
5. Karen, December 11, 2010 (12:29 p.m.), comment on T. Parker-Pope, "Pass the Pasta!" *Well* (blog), *New York Times*, December 10, 2010, http://well.blogs.nytimes.com/2010/12/10/pass-the-pasta/#comment-608687.
6. KevinJFUm, August 3, 2011 (9:26 p.m.), comment on topic "Are Meat Cravings Normal?" *Caveman Forum*, http://cavemanforum.com/diet-and-nutrition/are-meat-cravings-normal.
7. Cordain, *Paleo Diet*.
8. Vonderplanitz, *We Want to Live*.
9. Wolf, *Paleo Solution*.
10. Voegtlin, *Stone Age Diet*, 3.
11. Ibid., 1.
12. Ibid., 23–24.
13. Mallory, August 2, 2011 (5:09 p.m.), comment on topic "Do Carbs Make Your Nose Rounder?" *PaleoHacks*, http://paleohacks.com/questions/55442/do-carbs-make-your-nose-rounder#axzz20uvViyXl.
14. Evolution and Diseases of Modern Environments, at the Berlin Charité, October 2009.
15. Cordain, *Paleo Diet*, 3.
16. Ibid.
17. scott, October 19, 2010 (11:40 a.m.), comment on "The Stone Age Food Pyramid Included Flour Made from Wild Grains, *80beats* (blog), *Discover*, October 18, 2010, http://blogs.discovermagazine.com/80beats/2010/10/18/the-stone-age-food-pyramid-included-flour-made-from-wild-grains.

18. "Best Diets," *U.S. News & World Report*, accessed July 17, 2012, http://health.usnews.com/best-diet.
19. Avery Comarow, "Best Diets Methodology: How We Rated 25 Eating Plans," *U.S. News & World Report*, January 3, 2012, accessed July 17, 2012, http://health.usnews.com/best-diet/articles/2012/01/03/best-diets-methodology-how-we-rated-25-eating-plans.
20. "Paleo Diet," *U.S. News & World Report*, accessed July 17, 2012, http://health.usnews.com/best-diet/paleo-diet.
21. "Best Diets Overall," *U.S. News & World Report*, accessed July 17, 2012, http://health.usnews.com/best-diet/best-overall-diets?page=3.
22. "Caveman Fad Diet," *NHS Choices*, May 9, 2008, http://www.nhs.uk/news/2008/05May/Pages/Cavemanfaddiet.aspx.
23. Ibid.
24. Marlowe, "Hunter-Gatherers and Human Evolution."
25. Milton, "Hunter-Gatherer Diets" (2002).
26. Milton, "Hunter-Gatherer Diets" (2000).
27. Milton, "Hunter-Gatherer Diets" (2002).
28. Gibbons, "Evolutionary Theory of Dentistry."
29. Ibid.
30. Texas Parks and Wildlife Department, "Nutritional Data," accessed March 2011, http://www.tpwd.state.tx.us/exptexas/programs/wildgame/nutrition.
31. Pollan, "Breaking Ground."
32. Milton, "Hunter-Gatherer Diets" (2002).
33. Ibid.
34. Il Capo, August 15, 2011 (11:07 a.m.), comment on topic "New to This WOE but Need to Ask—Who Eats Potatoes?" *Caveman Forum*, http://cavemanforum.com/diet-and-nutrition/new-to-this-woe-but-need-to-ask-who-eats-potatoes.
35. Armelagos, "Omnivore's Dilemma."
36. Perry et al., "Diet and the Evolution of Human Amylase Gene."
37. Patin and Quintana-Murci, "Demeter's Legacy."
38. Perry et al., "Diet and the Evolution of Human Amylase Gene."
39. Oota et al., "Recent Origin and Cultural Reversion."
40. Luca et al., "Multiple Advantageous Amino Acid Variants."
41. Sabbagh et al., "Arylamine *N*-Acetyltransferase 2."
42. Luca, Perry, and Di Rienzo, "Evolutionary Adaptations."
43. Hehemann et al., "Transfer of Carbohydrate-Active Enzymes."
44. Kau et al., "Human Nutrition."

Chapter 6: Exercising the Paleofantasy

1. Goldstein, "New Age Cavemen and the City."
2. "How to Start," *CrossFit*, accessed July 24, 2012, http://www.crossfit.com/cf-info/start-how.html.
3. "What Is CrossFit?" *CrossFit*, accessed July 24, 2012, http://www.crossfit.com/cf-info/what-crossfit.html.
4. De Vany, "Art's . . . Essay on Evolutionary Fitness."
5. Ibid.
6. O'Keefe et al., "Achieving Hunter-Gatherer Fitness"; O'Keefe et al., "Exercise like a Hunter-Gatherer."
7. Ibid.
8. Owen et al., "Too Much Sitting."
9. Levine, "Nonexercise Activity Thermogenesis."
10. Owen et al., "Too Much Sitting."
11. Levine, "Nonexercise Activity Thermogenesis."
12. Ibid.
13. Levine et al., "Non-exercise Physical Activity."
14. Ibid.
15. Booth et al., "Reduced Physical Activity."
16. Booth et al., "Exercise and Gene Expression"; Booth et al., "Reduced Physical Activity."
17. De Vany, "Art's . . . Essay on Evolutionary Fitness."
18. O'Keefe et al., "Achieving Hunter-Gatherer Fitness"; O'Keefe et al., "Exercise like a Hunter-Gatherer."
19. De Vany, "Art's . . . Essay on Evolutionary Fitness."
20. Taleb, *Black Swan*, "Why I Do All This Walking."
21. Johnson, "Live Like a Caveman?"
22. Taleb, *Black Swan*, 26.
23. Ibid, 30.
24. Carrier, "Energetic Paradox."
25. McDougall, "Born to Run Marathons."
26. Carrier, "Energetic Paradox."
27. McDougall, *Born to Run*, 223.
28. Bramble and Lieberman, "Endurance Running."
29. Ibid.
30. Carrier, "Energetic Paradox."
31. Liebenberg, "Persistence Hunting."

32. Zorpette, "Louis Liebenberg."
33. McDougall, *Born to Run*, 239–240.
34. Bramble and Lieberman, "Endurance Running."
35. McDougall, "Born to Run Marathons."
36. Noakes and Spedding, "Run for Your Life."
37. Mark Sisson, "Who Is Grok?" *Mark's Daily Apple* (blog), accessed July 17, 2012. http://www.marksdailyapple.com/about-2/who-is-grok/#axzz1z7sfawh9.
38. Mark Sisson, "Did Humans Evolve to Be Long-Distance Runners?" *Mark's Daily Apple* (blog), April 21, 2009, http://www.marksdailyapple.com/did-humans-evolve-to-be-long-distance-runners/#axzz216klVrMo.
39. Taleb, *Black Swan*, "Why I Do All This Walking."
40. Bunn and Pickering, "Bovid Mortality Profiles."
41. Ibid.
42. Steudel-Numbers and Wall-Scheffler, "Optimal Running Speed."
43. Lieberman et al., "Evolution of Endurance Running."
44. Raichlen, Armstrong, and Lieberman, "Calcaneus Length."
45. Ruxton and Wilkinson, "Thermoregulation and Endurance Running."
46. Downey, "Lose Your Shoes."
47. Van Mechelen, "Running Injuries."
48. Macera et al., "Predicting Lower-Extremity Injuries."
49. "2011 Marathon, Half-Marathon and State of the Sport Results: Running USA's Annual Marathon Report," *Running USA*, March 16, 2011, http://www.runningusa.org/node/76115; "Running USA: Running Defies the Great Recession: Running USA's State of the Sport 2010—Part II," LetsRun.com, June 16, 2010, http://www.letsrun.com/2010/recessionproofrunning0617.php.
50. Newman, "Appealing to Runners."
51. Christopher McDougall, "The Barefoot Running Debate," *Christopher McDougall* (blog), accessed July 17, 2012, http://www.chrismcdougall.com/barefoot.html.
52. McDougall, *Born to Run*.
53. Parker-Pope, "Are Barefoot Shoes Really Better?"
54. Lieberman et al., "Foot Strike Patterns."
55. Christopher McDougall, "The Barefoot Running Debate," *Christopher McDougall* (blog), accessed July 17, 2012, http://www.chrismcdougall.com/barefoot.html.
56. Jenkins and Cauthon, "Barefoot Running Claims."

57. Colin John, June 8, 2011 (2:29 p.m.), comment on G. Reynolds, "Are We Built to Run Barefoot?" *Well* (blog), *New York Times*, June 8, 2011, http://well.blogs.nytimes.com/2011/06/08/are-we-built-to-run-barefoot/?comments#permid=91.
58. MarkRemy, June 8, 2011 (2:28 p.m.), comment on G. Reynolds, "Are We Built to Run Barefoot?" *Well* (blog), *New York Times*, June 8, 2011, http://well.blogs.nytimes.com/2011/06/08/are-we-built-to-run-barefoot/?comments#permid=87.
59. Downey, "Lose Your Shoes."
60. Berman and North, "Gene for Speed."
61. MacArthur et al., "Loss of ACTN3 Gene Function."
62. Ibid.
63. Ruiz et al., "Is There an Optimum?"
64. MacArthur, "ACTN3 Sports Gene Test."
65. Downey, "Lose Your Shoes."
66. Jacob, "Evolution and Tinkering."

Chapter 7: Paleofantasy Love

1. Shalit, "Is Infidelity Natural?"
2. vizirus, (date and time unknown), comment on W. Shalit, "Is Infidelity Natural? Ask the Apes," *CNN Opinion*, September 2, 2010. http:// www.cnn.com/2010/OPINION/09/02/shalit.infidelity/index.html#comment-74792430.
3. Ryan and Jethá, *Sex at Dawn*, 2.
4. celticcavegirl, April 21, 2011 (5:04 p.m.), reply #137 on topic "Involuntary Celibacy—The Underground Epidemic," *Caveman Forum*, http://cavemanforum.com/miscellaneous/involuntary-celibacy-the-underground-epidemic/137.
5. Il Capo, March 26, 2011 (12:52 p.m.), reply #3 on topic "Involuntary Celibacy—The Underground Epidemic," *Caveman Forum*, http://cavemanforum.com/miscellaneous/involuntary-celibacy-the-underground-epidemic/3.
6. smcdow, October 16, 2010 (4:36 p.m.), comment on topic "Sexual Habits of Our Ancestors," *PaleoHacks*, http://paleohacks.com/questions/12167/sexual-habits-of-our-ancestors#axzz21BBiSuft.
7. DAC, March 15, 2011 (12:37 p.m.), comment on topic "Has Going Paleo Made You Leave Behind Any Societal Norms Regarding Sexuality?" *PaleoHacks*,

http://paleohacks.com/questions/27619/has-going-paleo-made-you-leave-behind-any-societal-norms-regarding-sexuality#axzz21BBiSuft.

8. Trivers, "Parental Investment and Sexual Selection."
9. Ryan and Jethá, *Sex at Dawn*, 50.
10. De Waal, *Bonobo*, 134.
11. Ibid., 2.
12. Ryan and Jethá, *Sex at Dawn*.
13. Stanford, "Social Behavior of Chimpanzees and Bonobos."
14. Ibid.
15. Ibid.
16. Schultz, Opie, and Atkinson, "Stepwise Evolution."
17. Ingoldsby, "Marital Structure," 100.
18. Ryan, "Sex, Evolution, and the Case."
19. Fortunato, "Reconstructing the History of Marriage Strategies."
20. Hammer et al., "Ratio of Human X Chromosome."
21. Hager, "Sex and Gender in Paleoanthropology."
22. Zihlman, "Paleolithic Glass Ceiling."
23. Bird, "Cooperation and Conflict."
24. Ibid.
25. Codding, Bird, and Bird, "Provisioning Offspring and Others."
26. Hawkes, O'Connell, and Coxworth, "Family Provisioning."
27. Hawkes and Bird, "Showing Off."
28. Author conversation with Jane Lancaster, circa 1987.
29. Lovejoy, "Reexamining Human Origins."
30. De Waal, "Was 'Ardi' a Liberal?"
31. Holden and Mace, "Sexual Dimorphism in Stature."
32. Soulsbury, "Genetic Patterns of Paternity."
33. Ryan and Jethá, *Sex at Dawn*.

Chapter 8: The Paleofantasy Family

1. Iunabelle, June 22, 2011 (6:58 a.m.), comment on topic "Babies Crying Fixed with a Movement." *PaleoHacks*, http://paleohacks.com/questions/46489/babies-crying-fixed-with-a-movement#axzz22EKrng1y.
2. Hrdy, *Mothers and Others*, 69.
3. DeSilva, "Shift toward Birthing."

4. Rosemary Joyce, “Back-packing Mommas and Brainy Babies,” *What Makes Us Human—And One Percent Neanderthal* (blog), *Psychology Today*, October 27, 2010, http://www.psychologytoday.com/blog/what-makes-us-human/201010/back-packing-mommas-and-brainy-babies.
5. Gibbons, “Birth of Childhood.”
6. Blurton-Jones and Marlowe, “Selection for Delayed Maturity.”
7. Ibid.
8. Small, “Mother’s Little Helpers.”
9. Bogin, ““Evolutionary Hypotheses.”
10. Ibid.
11. Hrdy, *Langurs of Abu*.
12. Ibid.
13. Hrdy, *Mothers and Others*.
14. Ibid., 85.
15. Ibid., 109.
16. Ibid., 150.
17. Kruger and Konner, “Who Responds to Crying?”
18. Kramer, “Cooperative Breeding.”
19. Hrdy, *Mothers and Others*, 130.
20. Mace and Sear, “Are Humans Cooperative Breeders?”
21. Kramer, “Cooperative Breeding.”
22. Strassmann, “Cooperation and Competition.”
23. Hrdy, *Mothers and Others*, 128.
24. Gettler et al., “Longitudinal Evidence.”
25. Gray, “Descent of a Man’s Testosterone.”
26. Gettler, “Direct Male Care.”
27. Ibid.
28. Winking and Gurven, “Total Cost of Father Desertion.”
29. Hrdy, *Mothers and Others*.
30. Mace and Sear, “Are Humans Cooperative Breeders?”
31. Kaptijn et al., “How Grandparents Matter.”
32. Cant and Johnstone, “Reproductive Conflict.”
33. Hagen and Barrett, “Cooperative Breeding.”
34. Ibid.
35. McKenna, Ball, and Gettler, “Mother-Infant Cosleeping.”
36. Mansbach, *Go the F**k to Sleep*.
37. Small, *Our Babies, Ourselves*.

38. Ibid.
39. McKenna, Ball, and Gettler, "Mother-Infant Cosleeping."
40. McKenna, Ball, and Gettler, "Mother-Infant Cosleeping"; Mother-Baby Behavioral Sleep Laboratory, http://nd.edu/~jmckenn1/lab, accessed April 2012.
41. Gettler and McKenna, "Evolutionary Perspectives."
42. Small, *Our Babies, Ourselves*, 153.
43. Kruger and Konner, "Who Responds to Crying?"
44. "Babywearing in Church." TheBabyWearer.com, accessed April 2012. http://www.thebabywearer.com/index.php?page=bwchurch.

Chapter 9: Paleofantasy, in Sickness and in Health

1. markus, November 1, 2007, comment on M. Sisson, "The Biggest Myth about Cancer: That It Just 'Happens,'" *Mark's Daily Apple* (blog), October 31, 2007, http://www.marksdailyapple.com/cancer-myths-and-facts/#axzz22KPaVozt.
2. Mike OD, October 31, 2007, comment on M. Sisson, "The Biggest Myth about Cancer: That It Just 'Happens,'" *Mark's Daily Apple* (blog), October 31, 2007, http://www.marksdailyapple.com/cancer-myths-and-facts/#axzz22KPtoV3I.
3. Mennerat et al., "Intensive Farming."
4. Ibid.
5. Wells, *Pandora's Seed*, 89.
6. Gage, "Are Modern Environments Really Bad for Us?"
7. "Reconstructing Health and Disease in Europe: The Early Middle Ages through the Industrial Period." Poster presented at 78th Annual Meeting of the American Association of Physical Anthropologists, Chicago, IL, April 2009.
8. Domazet-Lošo and Tautz, "Ancient Evolutionary Origin of Genes."
9. Ibid.
10. Stephens et al., "Dating the Origin."
11. Galvani and Slatkin, "Evaluating Plague and Smallpox."
12. Galvani and Slatkin, "Intense Selection."
13. Novembre, Galvani, and Slatkin, "Geographic Spread."

14. Baron and Schembri-Wismayer, "Using the Distribution."
15. "Final 2011 West Nile Virus Human Infections in the United States." CDC, accessed July 2012, http://www.cdc.gov/ncidod/dvbid/westnile/surv&controlCaseCount11_detailed.htm.
16. Glass et al., "CCR5 Deficiency Increases Risk."
17. Donoghue, "Insights Gained from Palaeomicrobiology."
18. Ibid.
19. Barnes et al., "Ancient Urbanization Predicts."
20. "The History of Cancer," American Cancer Society, accessed April 2012. http://www.cancer.org/Cancer/CancerBasics/TheHistoryOfCancer.
21. "About the American Cancer Society," American Cancer Society, accessed April 2012, http://pressroom.cancer.org/index.php?s=43&item=52.
22. "Cancer," NewTreatments.org, accessed July 2012, http://www.newtreatments.org/cancer.
23. Ibid.
24. NewTreatments.org, http://www.newtreatments.org/index.
25. David and Zimmerman, "Cancer: An Old Disease."
26. R. David, in Coghlan, "Briefing: Cancer Is Not."
27. Deborah Mitchell, "Mummies Don't Lie: Cancer Is Modern and Man Made," *EmaxHealth*, October 15, 2010, http://www.emaxhealth.com/1275/mummies-dont-lie-cancer-modern-and-man-made.
28. Richard Alleyne, "Cancer Caused by Modern Man as It Was Virtually Non-existent in Ancient World," *Telegraph*, October 14, 2010, http://www.telegraph.co.uk/health/healthnews/8064554/Cancer-caused-by-modern-man-as-it-was-virtually-non-existent-in-ancient-world.html.
29. "Report to the Nation Finds Continuing Declines in Cancer Death Rates since the Early 1990s," NCI Press Release, National Cancer Institute, accessed August 2012, http://www.cancer.gov/newscenter/pressreleases/2012/ReportNationRelease2012.
30. Waldron, "What Was the Prevalence?"
31. Nerlich and Bachmeier, "Paleopathology of Malignant Tumours."
32. Coghlan, "Briefing: Cancer Is Not."
33. Finch, "Evolution of the Human Lifespan."
34. Greaves, "Darwinian Medicine."
35. Crespi, "Emergence of Human-Evolutionary Medical Genomics."
36. Ibid.

Chapter 10: Are We Still Evolving? A Tale of Genes, Altitude, and Earwax

1. Stiffler, January 7, 2010 (12:08 p.m.), comment on C. Zimmer, "The Origin of the Future: Death by Mutation?" *The Loom* (blog), *Discover Magazine*, January 7, 2010, http://blogs.discovermagazine.com/loom/2010/01/07/the-origin-of-the-future-death-by-mutation.
2. klcarbaugh, October 31, 2009 (12:22 p.m.), comment on topic "Human Evolution Speeding Up," *Caveman Forum*, http://cavemanforum.com/research/human-evolution-speeding-up/msg16332/#msg16332.
3. Destor, October 4, 2010 (12:04 p.m.), comment on topic "Rapid Evolutionary Adaptations," *Caveman Forum*, http://cavemanforum.com/diet-and-nutrition/rapid-evolutionary-adaptations/msg38901/#msg38901.
4. BragonDorn, January 7, 2012 (5:22 p.m.), comment on Heskew, "Are Humans Still Evolving?" *Sports Abode* (blog), January 7, 2012, http://www.thesportsabode.com/2012/01/are-humans-still-evolving.html.
5. Alfred Lord Tennyson, *In Memoriam A. H. H.*
6. Greg, "On the Failure."
7. Tait, "Has the Law of Natural Selection?"
8. Darwin, *Descent of Man*, 501.
9. Shepherd, "Lawson Tait."
10. McKie, "Is Human Evolution Finally Over?"
11. Stock, "Are Humans Still Evolving?"
12. Meredith Small, quoted in "Ask the Experts: Are Human Beings Still Evolving? It Would Seem That Evolution Is Impossible Now That the Ability to Reproduce Is Essentially Universally Available. Are We Nevertheless Changing as a Species?" *Scientific American*, October 21, 1999, http://www.scientificamerican.com/article.cfm?id=are-human-beings-still-ev.
13. Hawks, "Human Evolution Stopping?"
14. Ibid.
15. Ibid.
16. Ibid.
17. Balter, "Are Humans Still Evolving?"
18. Byars et al., "Natural Selection."
19. Ibid.
20. Stearns et al., "Measuring Selection."
21. Milot et al., "Evidence for Evolution."
22. Stearns et al., "Measuring Selection."

23. Ibid.
24. Ibid.
25. Yi et al., "Sequencing of 50 Human Exomes."
26. Simonson et al., "Genetic Evidence."
27. Storz, "Genes for High Altitudes."
28. Williamson et al., "Localizing Recent Adaptive Evolution."
29. Ibid.
30. Hernandez et al., "Classic Selective Sweeps Were Rare."
31. Yoshiura et al., "SNP in the *ABCC11* Gene."
32. Ward, "What Will Become of *Homo sapiens*?"
33. Allen, "Why Humans Are Crazy for Crispy."
34. Graves, "Feel as Happy as a Pig in Mud!"
35. Michelle, January 4, 2012, comment on M. Sisson, "How Much Have Human Dietary Requirements Evolved in the Last 10,000 Years?" *Mark's Daily Apple* (blog), January 4, 2012, http://www.marksdailyapple.com/are-humans-still-evolving/#axzz22WKGUl1R.

Bibliography

Introduction

Balter, M. "How Human Intelligence Evolved—Is It Science or 'Paleofantasy'?" *Science* 319 (2008): 1028.

Cochran, G., and Harpending, H. *The 10,000 Year Explosion: How Civilization Accelerated Human Evolution*. New York: Basic Books, 2009.

Coyne, J. A. *Why Evolution Is True*. New York: Penguin, 2010.

Holland, J. "The Cavewoman's Guide to Good Health." *Glamour*, September 2010. http://www.glamour.com/health-fitness/2010/09/the-cavewomans-guide-to-good-health.

McKie, R. "Is Human Evolution Finally Over? *Observer*, February 2, 2002.

Shubin, N. *Your Inner Fish*. New York: Vintage, 2009.

Simpson, G. G. *Tempo and Mode in Evolution*. New York: Columbia University Press, 1944.

Tinghitella, R. M., and Zuk, M. "Asymmetric Mating Preferences Accommodated the Rapid Evolutionary Loss of a Sexual Signal." *Evolution* 63 (2009): 2087–98.

Wade, N. *Before the Dawn: Recovering the Lost History of Our Ancestors*. New York: Penguin, 2007.

Zuk, M., Rotenberry, J. T., and Tinghitella, R. M. "Silent Night: Adaptive Disappearance of a Sexual Signal in a Parasitized Population of Field Crickets. *Biology Letters* 2 (2006): 521–24.

Chapter 1: Cavemen in Condos

Alleyne, R. "Neanderthals Really Were Sex-Obsessed Thugs." *Telegraph*, November 3, 2010. http://www.telegraph.co.uk/science/evolution/8104939/Neanderthals-really-were-sex-obsessed-thugs.html.

Bower, B. "The Strange Case of the Tasaday." *Science News* 135 (1989): 280–83.

Boyd, R., and Silk, J. B. *How Humans Evolved*. 5th ed. New York: Norton, 2009.

———. *How Humans Evolved*. 6th ed. New York: Norton, 2012.

Brown, D. E. *Human Universals*. New York: McGraw-Hill, 1991.

"Cavemen Randier Than People Today." *Mirror*, November 3, 2010. http://www.mirror.co.uk/news/uk-news/cavemen-randier-than-people-today-259336.

"Diet Fad from the Stone Age." *Sydney Morning Herald*, February 8, 2010. http://www.smh.com.au/lifestyle/wellbeing/diet-fad-from-the-stone-age-20100208-nmol.html.

Enard, W., Przeworski, M., Fisher, S. E., Lai, C. S. L., Wiebe, V., Kitano, T., Monaco, A. P., and Paabo, S. "Molecular Evolution of FOXP2, a Gene Involved in Speech and Language." *Nature* 418 (2002): 869–72. doi:10.1038/nature01025.

"Evolution: Neanderthals Matured Fast." *Nature* 468 (2010): 478. doi:10.1038/468478c.

Gibbons, A. "Lucy's Toolkit? Old Bones May Show Earliest Evidence of Tool Use." *Science* 329 (2010): 738–39.

Glendinning, L., and agencies. "Neanderthals: Not Stupid, Just Different." *Guardian*, August 26, 2008. http://www.guardian.co.uk/science/2008/aug/26/evolution.

Goldstein, J. "The New Age Cavemen and the City." *New York Times*, January 8, 2010.

Green, R. E., Krause, J., Briggs, A. W., Maricic, T., Stenzel, U., Kircher, M., Patterson, N., et al. "A Draft Sequence of the Neandertal Genome." *Science* 328 (2010): 710–22. doi:10.1126/science.1188021.

Hawks, J. "The Cavemen Are Happy in the Modern World." *John Hawks Weblog* (blog), January 11, 2010. http://johnhawks.net/weblog/topics/humor/caveman-diet-nytimes-2010.html.

Hawks, J. "'Naughty Neandertals' Did What?" *John Hawks Weblog* (blog), September 24, 2010. http://johnhawks.net/weblog/reviews/neandertals/development/digit-ratio-nelson-neandertal-2009.html.

———. "Neandertal Stories on Parade." *John Hawks Weblog* (blog), December 4, 2010. http://johnhawks.net/weblog/reviews/neandertals/symbolism/mckie-neandertal-story-2010.html.

———. "Ozzy Osbourne, Archaic Human." *John Hawks Weblog* (blog), October 25, 2010. http://johnhawks.net/node/14969.

Jacobsen, H. "Essential Caveman Lifestyle and Environmental Changes: Maintaining a Caveman Body Is a Four Legged Stool." Diabetes Cure 101, March 6, 2006, http://diabetescure101.com/cavemansbody.shtml.

Joyce, R. "Do Our Ancestors Walk among Us?" *What Makes Us Human* (blog), *Psychology Today*, September 22, 2010. http://www.psychologytoday.com/blog/what-makes-us-human/201009/do-our-ancestors-walk-among-us.

———. "Fingering Neanderthal Sexuality." *What Makes Us Human* (blog), *Psychology Today*, November 4, 2010. http://www.psychologytoday.com/node/50068.

Keim, B. "Neanderthals Not Dumb, but Made Dull Gadgets." *Wired Science* (blog), *Wired*, August 26, 2008. http://www.wired.com/wiredscience/2008/08/neanderthals-no.

Kratochvíl, L., and Flegr, J. "Differences in the 2nd to 4th Digit Length Ratio in Humans Reflect Shifts along the Common Allometric Line." *Biology Letters* 5 (2009): 643–46.

Manning, J. T. *Digit Ratio: A Pointer to Fertility, Behavior and Health*. New Brunswick, NJ: Rutgers University Press, 2002.

Martin, A. "We've All Suspected, Now It's Official: Ozzy Osbourne IS a Neanderthal." *Daily Mail*, October 25, 2010. http://www.dailymail.co.uk/sciencetech/article-1323455/Weve-suspected-official-Ozzy-Osbourne-IS-Neanderthal.html.

Milius, S. "Tapeworms Tell Tales of Deeper Human Past." *Science News*, April 7, 2001. http://findarticles.com/p/articles/mi_m1200/is_14_159/ai_104730217.

"Neanderthals Had a Naughty Sex Life, Unusual Study Suggests." Agence France Press, November 2, 2010. http://www.google.com/hostednews/afp/article/ALeqM5j30jS1P2CP41qjX2cq_k_ANwgmDA?docId=CNG.d38937404101e4d7e98cb91a23a3c053.41.

Nelson, E., Rolian, C., Cashmore, L., and Shultz, S. "Digit Ratios Predict Polygyny in Early Apes, *Ardipithecus*, Neanderthals and Early Modern Humans but Not in *Australopithecus*. *Proceedings of the Royal Society. B* 278 (2010): 1556–63. doi:10.1098/rspb.2010.1740.

Simopoulos, A. P., and Robinson, J. *The Omega Diet: The Lifesaving Nutritional Program Based on the Best of the Mediterranean Diets*. New York: Harper Paperbacks, 1999.

Smith, T. M., Tafforeau, P., Reid, D. J., Pouech, J., Lazzari, V., Zermeno, J. P., Guatelli-Steinberg, D., et al. "Dental Evidence for Ontogenetic Differences between Modern Humans and Neanderthals." Preprint, submitted to *Proceedings of the National Academy of Sciences of the USA* July 26, 2010. http://www.pnas.org/content/early/2010/11/08/1010906107.

Sponheimer, M., and Lee-Thorp, J. A. "Isotopic Evidence for the Diet of an Early Hominid, *Australopithecus africanus*." *Science* 283 (1999): 368–70.

"Stone Me—He's Smart, He's Tough, and He's Equal to Any Homo sapiens." Scotsman.com, August 25, 2008. http://www.scotsman.com/news/uk/stone-me-he-s-smart-he-s-tough-and-he-s-equal-to-any-homo-sapiens-1-1087209.

Wade, N. *Before the Dawn: Recovering the Lost History of Our Ancestors*. New York: Penguin, 2007.

White, T. D., Asfaw, B., Beyene, Y., Haile-Selassie, Y., Lovejoy, C. O., Suwa, G., and WoldeGabriel, G. "*Ardipithecus ramidus* and the Paleobiology of Early Hominids." *Science* 326 (2009): 75–86.

White, T. H. "Hominid Paleobiology: How Has Darwin Done?" In *Evolution since Darwin: The First 150 Years*, edited by M. A. Bell, D. J. Futuyma, W. F. Eanes, and J. S. Levinton, 519–60. Sunderland, MA: Sinauer, 2010.

Wrangham, R. *Catching Fire: How Cooking Made Us Human*. New York: Basic Books, 2009.

Wrangham, R., and Peterson, D. *Demonic Males: Apes and the Origins of Human Violence*. Boston: Houghton Mifflin, 1996.

Wrangham, R., and Pilbeam, D. "African Apes as Time Machines." In *The Human Evolution Source Book*, 2nd ed., edited by R. L. Ciochon and J. G. Fleagle, 33–38. Upper Saddle River, NJ: Pearson Prentice Hall, 2006.

Zihlman, A. "The Paleolithic Glass Ceiling: Women in Human Evolution." In *Women in Human Evolution*, edited by L. D. Hager, 91–114. London: Routledge, 1997.

Zimmer, C. *Smithsonian Intimate Guide to Human Origins*. New York: Harper Perennial, 2007.

Chapter 2: Are We Stuck?

Bowles, S. "Did Warfare among Ancestral Hunter-Gatherers Affect the Evolution of Human Social Behaviors?" *Science* 324 (2009): 1293–98.

Britten, R. J. "Divergence between Samples of Chimpanzee and Human DNA Sequences Is 5%, Counting Indels." *Proceedings of the National Academy of Sciences of the USA* 99 (2002): 13633–35.

Carroll, S. B. "Genetics and the Making of *Homo sapiens*." *Nature* 422 (2003): 849–57.

Cochran, G., and Harpending, H. *The 10,000 Year Explosion: How Civilization Accelerated Human Evolution*. New York: Basic Books, 2009.

Cordain, L. *The Paleo Diet: Lose Weight and Get Healthy by Eating the Foods You Were Designed to Eat*. New York: Wiley, 2001.

Cosmides, L., and Tooby, J. "Evolutionary Psychology: A Primer." Center for Evolutionary Psychology, last modified January 13, 1997, http://www.psych.ucsb.edu/research/cep/primer.html.

Denham, T. P., Iriarte, J., and Vrydaghs, L., eds. *Rethinking Agriculture: Archaeological and Ethnoarchaeological Perspectives*. Walnut Creek, CA: Left Coast Press, 2009.

Diamond, J. "The Worst Mistake in the History of the Human Race." *Discover*, May 1987.

Feeney, J. "Agriculture: Ending the World as We Know It." *Zephyr*, August–September 2010.

Gage, T. B. "Are Modern Environments Really Bad for Us?: Revisiting the Demographic and Epidemiologic Transitions." *Yearbook of Physical Anthropology* 48 (2005): 96–117.

Harris, D. R. "Agriculture, Cultivation, and Domestication: Exploring the Conceptual Framework of Early Food Production." In *Rethinking Agriculture: Archaeological and Ethnoarchaeological Perspectives*, edited by T. P. Denham, J. Iriarte, and L. Vrydaghs, 16–35. Walnut Creek, CA: Left Coast Press, 2009.

Hawks, J. "Why Human Evolution Accelerated." *John Hawks Weblog* (blog), December 12, 2007. http://johnhawks.net/weblog/topics/evolution/selection/acceleration/accel_story_2007.html.

Irons, W. "Adaptively Relevant Environments versus the Environment of Evolutionary Adaptedness." *Evolutionary Anthropology* 6 (1998): 194–204.

Kaplan, H. S., Hill, K., Lancaster, J. B., and Hurtado, A. M. "A Theory of Human Life History Evolution: Diet, Intelligence, and Longevity." *Evolutionary Anthropology* 9 (2000): 156–85.

Lee, R. B. *The !Kung San: Men, Women and Work in a Foraging Society*. Cambridge: Cambridge University Press, 1979.

Marks, J. *What It Means to Be 98% Chimpanzee: Apes, People, and Their Genes*. Berkeley: University of California Press, 2002.

O'Connell, S. "Is Farming the Root of All Evil?" *Telegraph*, June 23, 2009. http://www.telegraph.co.uk/science/science-news/5604296/Is-farming-the-root-of-all-evil.html.

Powell, A., Shennan, S., and Thomas, M. G. "Late Pleistocene Demography and the Appearance of Modern Human Behavior." *Science* 324 (2009): 1298–1301.

Radosavljevic, Z. "Stone Age Sights, Sounds, Smells at Croat Museum." *Reuters*, March 1, 2010. http://www.reuters.com/article/2010/03/01/us-neanderthal-croatia-museum-idUSTRE6202EW20100301.

Singer, P. "Is Violence History?" *New York Times*, October 6, 2011. http://www.nytimes.com/2011/10/09/books/review/the-better-angels-of-our-nature-by-steven-pinker-book-review.html?_r=2&pagewanted=all.

Strassmann, B. I., and Dunbar, R. I. M. "Human Evolution and Disease: Putting the Stone Age in Perspective." In *Evolution in Health and Disease*, edited by S. C. Stearns, 91–101. Oxford: Oxford University Press, 1999.

Tooby, J., and Cosmides, L. "The Past Explains the Present: Emotional Adaptations and the Structure of Ancestral Environments." *Ethology and Sociobiology* 11 (1990): 375–424.

Wells, S. *Pandora's Seed: The Unforeseen Cost of Civilization*. New York: Random House, 2010.

Wright, S. "The Roles of Mutation, Inbreeding, Crossbreeding and Selection in Evolution." In *Proceedings of the Sixth International Congress of Genetics, Ithaca, New York, 1932, Vol. 1: Transactions and General Addresses*, edited by Donald F. Jones, 356–66. 1932.

Chapter 3: Crickets, Sparrows, and Darwins—or, Evolution before Our Eyes

Allendorf, F. W., and Hard, J. L. "Human-Induced Evolution Caused by Unnatural Selection through Harvest of Wild Animals." *Proceedings of the National Academy of Sciences of the USA* 106 (2009): 9987–94.

Barrett, R. D. H., Paccard, A., Healy, T. M., Bergek, S., Schulte, P. M., Schluter, D., and Rogers, S. M. "Rapid Evolution of Cold Tolerance in Stickleback." *Proceedings of the Royal Society B* 278 (2010): 233–38.

Berthold, P., Helbig, A. J., Mohr, G., and Querner, U. "Rapid Microevolution of Migratory Behaviour in a Wild Bird Species." *Nature* 360 (1992): 668–70.

Bumpus, H. C. "The Elimination of the Unfit as Illustrated by the Introduced House Sparrow, *Passer domesticus*." *Biological Lectures Delivered at the Marine Biological Laboratory at Woods Hole* (1899), 209–25.

Carroll, S. P. "Facing Change: Forms and Foundations of Contemporary Adaptation to Biotic Invasions." *Molecular Ecology* 17 (2008): 361–72.

Carroll, S. P., Hendry, A. P., Reznick, D. N., and Fox, C. W. "Evolution on Ecological Time-Scales." *Functional Ecology* 21 (2007): 387–93.

Carroll, S. P., Loye, J. E., Dingle, H., Mathieson, M., Famula, T. R., and Zalucki,

M. P. "And the Beak Shall Inherit—Evolution in Response to Invasion." *Ecology Letters* 8 (2005): 944–51.

Coltman, D. W., O'Donoghue, P., Jorgenson, J. T., Strobeck, C., Festa-Bianchet, M., and Hogg, J. T. "Undesirable Evolutionary Consequences of Trophy Hunting." *Nature* 426 (2003): 655–58.

Eldridge, W. H., Hard, J. J., and Naish, K. A. "Simulating Fishery-Induced Evolution in Chinook Salmon: The Role of Gear, Location, and Genetic Correlation among Traits." *Ecological Applications* 20 (2010): 1936–48.

Elmer, K. R., Lehtonen, T. K., Kautt, A. F., Harrod, C., and Meyer, A. "Rapid Sympatric Ecological Differentiation of Crater Lake Cichlid Fishes within Historic Times." *BMC Biology* 8 (2010): 60.

Gingerich, P. D. "Quantification and Comparison of Evolutionary Rates." *American Journal of Science*, 293 (1993): 453–78.

———."Rates of Evolution: Effects of Time and Temporal Scaling." *Science* 222 (1983): 159–61.

Grant, P. R., and Grant, B. R. "Unpredictable Evolution in a 30-Year Study of Darwin's Finches." *Science* 296 (2002): 707–11.

Haldane, J. B. S. "Suggestions as to Quantitative Measurement of Rates of Evolution." *Evolution* 3 (1949): 51–56.

Halfwerk, W., Holleman, L., Lessells, K., and Slabbekoorn, H. "Negative Impact of Traffic Noise on Avian Reproductive Success." *Journal of Applied Ecology* 48 (2011): 210–19.

Hendry, A. P., and Kinnison, M. T. "Perspective: The Pace of Modern Life: Measuring Rates of Contemporary Microevolution." *Evolution* 53 (1999): 1637–53.

Phillips, B. L., Brown, G. P., and Shine, R. "Evolutionarily Accelerated Invasions: The Rate of Dispersal Evolves Upwards during the Range Advance of Cane Toads." *Journal of Evolutionary Biology* 23 (2010): 2595–601.

Phillips, B. L., and Shine, R. "Adapting to an Invasive Species: Toxic Cane Toads Induce Morphological Change in Australian Snakes." *Proceedings of the National Academy of Sciences of the USA* 101 (2004): 17150–55.

Reznick, D. N., Ghalambor, C. K., and Crooks, K. "Experimental Studies of Evolution in Guppies: A Model for Understanding the Evolutionary Consequences of Predator Removal in Natural Communities." *Molecular Ecology* 17 (2008): 97–107.

Reznick, D. N., Shaw, F. H., Rodd, F. H., and Shaw, R. G. "Evaluation of the Rate of Evolution in Natural Populations of Guppies (*Poecilia reticulata*)." *Science* 275 (1997): 1934–37.

Rudolf, J. C. "Speedy Evolution, Indeed." *New York Times*, February 18, 2011.

Schoener, T. W. "The Newest Synthesis: Understanding the Interplay of Evolutionary and Ecological Dynamics." *Science* 331 (2011): 426.

Seger, J. "El Niño and Darwin's Finches." *Nature* 327 (1987): 461.

Wirgin, I., Roy, N. K., Loftus, M., Chambers, R. C., Franks, D. G., and Hahn, M. E. "Mechanistic Basis of Resistance to PCBs in Atlantic Tomcod from the Hudson River." *Science* 331 (2011): 1322–25.

Wolak, M. E., Gilchrist, G. W., Ruzicka, V. A., Nally, D. M., and Chambers, R. M. "A Contemporary, Sex-Limited Change in Body Size of an Estuarine Turtle in Response to Commercial Fishing." *Conservation Biology* 24 (2010): 1268–77.

Chapter 4: The Perfect Paleofantasy Diet: Milk

Anderson, B., and Vullo, C. "Did Malaria Select for Primary Adult Lactase Deficiency?" *Gut* 35 (1994): 1487–89.

Beja-Pereira, A., Luikart, G., England, P. R., Bradley, D. G., Jann, O. C., Bertorelle, G., Chamberlain, A. T., et al. "Gene-Culture Coevolution between Cattle Milk Protein Genes and Human Lactase Genes." *Nature Genetics* 35 (2003): 311–13.

Bloom, G., and Sherman, P. W. "Dairying Barriers Affect the Distribution of Lactose Malabsorption." *Evolution and Human Behavior* 26 (2005): 301.e1–33.

Boyd, R., and Silk, J. B. *How Humans Evolved.* 6th ed. New York: Norton, 2012.

Gerbault, P., Liebert, A., Itan, Y., Powell, A., Currat, M., Burger, J., Swallow, D. M., and Thomas, M. G. "Evolution of Lactase Persistence: An Example of Human Niche Construction." *Philosophical Transactions of the Royal Society B* 366 (2011): 863–77.

Gerbault, P., Moret, C., Currat, M., and Sanchez-Mazas, A. "Impact of Selection and Demography on the Diffusion of Lactase Persistence." *PLoS ONE* 4 (2009): e6369.

Ingram, C. J. E., Mulcare, C. A., Itan, Y., Thomas, M. G., and Swallow, D. M. "Lactose Digestion and the Evolutionary Genetics of Lactase Persistence." *Human Genetics* 124 (2009): 579–91.

Itan, Y., Jones, B. L., Ingram, C. J. E., Swallow, D. M., and Thomas, M. G. "A Worldwide Correlation of Lactase Persistence Phenotype and Genotypes." *BMC Evolutionary Biology* 10 (2010): 36.

Kiple, K. F., and Ornelas, K. C., eds. *The Cambridge World History of Food.* Cambridge: Cambridge University Press, 2000.

Scheindlin, B. "Lactose Intolerance and Evolution: No Use Crying over Undigested Milk." *Gastronomica: The Journal of Food and Culture* 7 (2007): 59–63.

Schiebinger, L. "Why Mammals Are Called Mammals: Gender Politics in Eighteenth-Century Natural History." *American Historical Review* 98 (1993): 382–411.

Tishkoff, S. A., Reed, F. A., Ranciaro, A., Voight, B. F., Babbitt, C. C., Silverman, J. S., Powell, K., et al. "Convergent Adaptation of Human Lactase Persistence in Africa and Europe." *Nature Genetics* 39 (2007): 31–40.

Chapter 5: The Perfect Paleofantasy Diet: Meat, Grains, and Cooking

Armelagos, G. J. "The Omnivore's Dilemma: The Evolution of the Brain and the Determinants of Food Choice." *Journal of Anthropological Research* 66 (2010): 161–86.

Babbitt, C. C., Warner, L. R., Fedrigo, O., Wall, C. E., and Wray, G. E. "Genomic Signatures of Diet-Related Shifts during Human Origins." *Proceedings of the Royal Society B* 278 (2011): 961–69.

Cordain, L. *The Paleo Diet: Lose Weight and Get Healthy by Eating the Foods You Were Designed to Eat*. New York: Wiley, 2001.

Cordain, L., Miller, J. B., Eaton, S. B., and Mann, N. "Macronutrient Estimations in Hunter-Gatherer Diets." *American Journal of Clinical Nutrition* 72 (2000): 1589–90.

———."Reply to SC Cunnane." *American Journal of Clinical Nutrition* 72 (2000): 1585–86.

Cunnane, S. C. "Hunter-Gatherer Diets—A Shore-Based Perspective." *American Journal of Clinical Nutrition* 72 (2000): 1583–84.

Gibbons, A. "An Evolutionary Theory of Dentistry." *Science* 336 (2012): 973–75.

Hehemann, J.-H., Correc, G., Barbeyron, T., Helbert, W., Czjzek, M., and Michel, G. "Transfer of Carbohydrate-Active Enzymes from Marine Bacteria to Japanese Gut Microbiota." *Nature* 464 (2010): 908–14.

Henry, A. G., Brooks, A. S., and Piperno, D. R. "Microfossils in Calculus Demonstrate Consumption of Plants and Cooked Foods in Neanderthal Diets (Shanidar III, Iraq; Spy I and II, Belgium)." *Proceedings of the National Academy of Sciences of the USA* 108 (2010): 486–91. doi:10.1073/pnas.1016868108.

Henry, A. G., Ungar, P. S., Passey, B. H., Sponheimer, M., Rossouw, L., Bamford, M., Sandberg, P., de Ruiter, D. J., and Berger, L. "The Diet of *Australopithecus sediba*." *Nature* 487 (2012): 90–93.

Hirst, K. K. "Grinding Flour in the Upper Paleolithic." About.com, October 18, 2010. http://archaeology.about.com/b/2010/10/18/grinding-flour-in-the-upper-paleolithic.htm.

Hobson, K. "Paleo Diet: Can Our Caveman Ancestors Teach Us the Best Modern Diet?" *U.S. News & World Report*, April 28, 2009.

Jew, S., AbuMweis, S. S., and Jones, P. J. H. "Evolution of the Human Diet: Linking Our Ancestral Diet to Modern Functional Foods as a Means of Chronic Disease Prevention." *Journal of Medicinal Food* 12 (2009): 925–34.

Kau, A. L., Ahern, P. P., Griffin, N. W., Goodman, A. L., and Gordon, J. I. "Human Nutrition, the Gut Microbiome and the Immune System." *Nature* 474 (2011): 327–36.

Kleim, B. "Ancient Grains Show Paleolithic Diet Was More Than Meat." *Wired Science* (blog), *Wired*, October 18, 2010. http://www.wired.com/wiredscience/2010/10/revised-paleolithic-diet.

Lawler, A. "Early Farmers Went Heavy on the Starch." *Science* 332 (2011): 416–17.

Luca, F., Bubba, G., Basile, M., Brdicka, R., Michalodimitrakis, E., Rickards, O., Vershubsky, G., Quintana-Murci, L., Kozlov, A. I., and Novelletto, A. "Multiple Advantageous Amino Acid Variants in the NAT2 Gene in Human Populations." *PLoS ONE* 3 (2008): e3136.

Luca, F., Perry, G. H., and Di Rienzo, A. "Evolutionary Adaptations to Dietary Changes." *Annual Reviews in Nutrition* 30 (2010): 291–314.

Marlowe, F. W. "Hunter-Gatherers and Human Evolution." *Evolutionary Anthropology* 14 (2005): 54–67.

Milton, K. "Hunter-Gatherer Diets—A Different Perspective." *American Journal of Clinical Nutrition* 71 (2000): 665–67.

———. "Hunter-Gatherer Diets: Wild Foods Signal Relief from Diseases of Affluence." In *Human Diet: Its Origin and Evolution*, edited by P. S. Ungar and M. F. Teaford, 111–22. Westport, CT: Bergen and Garvey, 2002.

———. "Reply to L. Cordain et al." *American Journal of Clinical Nutrition* 72 (2000): 1590–92.

———. "Reply to SC Cunnane." *American Journal of Clinical Nutrition* 72 (2000): 1585–86.

Minogue, K. "The Cavemen's Complex Kitchen." *Science Now*, October 18, 2010.

Oota, H., Pakendorf, B., Weiss, G., von Haeseler, A., Pookajorn, S., Settheetham-Ishida, W., Tiwawech, D., Ishida, T., and Stoneking, M. "Recent Origin and Cultural Reversion of a Hunter–Gatherer Group." *PLoS Biology* 3 (2005): e71.

Patin, E., and Quintana-Murci, L. "Demeter's Legacy: Rapid Changes to Our Genome Imposed by Diet." *Trends in Ecology and Evolution* 23 (2008): 56–69.

Perkes, C. "A Diet Plate Right Out of History." *Orange County Register*, June 14, 2011.

Perry, G. H., Dominy, N. J., Claw, K. G., Lee, A. S., Fiegler, H., Redon, R., Werner, J., et al. "Diet and the Evolution of Human Amylase Gene Copy Number Variation." *Nature Genetics* 39 (2007): 1256–60.

Pollan, M. "Breaking Ground; the Call of the Wild Apple." *New York Times*, November 5, 1998. http://www.nytimes.com/1998/11/05/garden/breaking-ground-the-call-of-the-wild-apple.html?pagewanted=all&src=pm.

Reuters. "Paleolithic Humans Had Bread Along with Their Meat." *New York Times*, October 18, 2010.

Revedin, A., Aranguren, B., Becattini, R., Longo, L., Marconi, E., Lippi, M. M., Skakun, N., Sinitsyn, A., Spiridonova, E., and Svoboda, J. "Thirty Thousand-Year-Old Evidence of Plant Food Processing." *Proceedings of the National Academy of Sciences of the USA* 107 (2010): 18815–19.

Sabbagh, A., Darlu, P., Crouau-Roy, B., and Poloni, E. S. "Arylamine *N*-Acetyltransferase 2 (*NAT2*) Genetic Diversity and Traditional Subsistence: A Worldwide Population Survey." *PLoS ONE* 6 (2011): e18507.

Schoeninger, M. J. "The Ancestral Dinner Table." *Nature* 487 (2012): 42–43.

Sponheimer, M., and Lee-Thorp, J. A. "Isotopic Evidence for the Diet of an Early Hominid, *Australopithecus africanus*." *Science* 283 (1999): 368–70.

"The Stone Age Food Pyramid Included Flour Made from Wild Grains." *80beats* (blog), *Discover*, October 18, 2010. http://blogs.discovermagazine.com/80beats/2010/10/18/the-stone-age-food-pyramid-included-flour-made-from-wild-grains.

Viegas, J. "Cavemen Ground Flour, Prepped Veggies." ABC Science, October 19, 2010. http://www.abc.net.au/science/articles/2010/10/19/3042264.htm.

Voegtlin, W. *The Stone Age Diet*. New York: Vantage, 1975.

Vonderplanitz, A. *We Want to Live*. Santa Monica, CA: Carnelian Bay Castle Press, 2005.

Wolf, R. *The Paleo Solution: The Original Human Diet*. Victory Belt Publishing, 2010.

Chapter 6: Exercising the Paleofantasy

Berman, Y., and North, K. N. "A Gene for Speed: The Emerging Role of a-Actinin-3 in Muscle Metabolism." *Physiology* 25 (2010): 250–59.

Booth, F. W., Chakravarthy, M. V., and Spangenburg, E. E. "Exercise and Gene Expression: Physiological Regulation of the Human Genome through Physical Activity." *Journal of Physiology* 543(Pt.2) (2002): 399–411.

Booth, F. W., Laye, M. J., Lees, S. J., Rector, R. S., and Thyfault, J. P. "Reduced Physical Activity and Risk of Chronic Disease: The Biology behind the Consequences." *European Journal of Applied Physiology* 102 (2008): 381–90.

Bramble, D. M., and Lieberman, D. E. "Endurance Running and the Evolution of *Homo*." *Nature* 432 (2004): 345–52.

Bunn, H. T., and Pickering, T. R. "Bovid Mortality Profiles in Paleoecological Context Falsify Hypotheses of Endurance Running–Hunting and Passive Scavenging by Early Pleistocene Hominins." *Quaternary Research* 74 (2010): 395–404.

Carrier, D. R. "The Energetic Paradox of Human Running and Hominid Evolution." *Current Anthropology* 25 (1984): 483–95.

De Vany, A. "Art's (Slightly Edited Original) Essay on Evolutionary Fitness." *Arthur De Vany's Evolutionary Fitness* (blog), October 22, 2010. http://www.arthurdevany.com/articles/20101022.

Downey, G. "Lose Your Shoes: Is Barefoot Better?" *Neuroanthropology* (blog), July 26, 2009. http://neuroanthropology.net/2009/07/26/lose-your-shoes-is-barefoot-better.

Goldstein, J. "The New Age Cavemen and the City." *New York Times*, January 8, 2010.

Jacob, F. "Evolution and Tinkering." *Science* 196 (1977): 1161–66.

Jenkins, D. W., and Cauthon, D. J. "Barefoot Running Claims and Controversies: A Review of the Literature." *Journal of the American Podiatric Medical Association* 101 (2011): 231–46.

Johnson, D. P. "Live Like a Caveman?" *D Patrick Johnson* (blog), January 25, 2010. http://dpatrickjohnson.wordpress.com/2010/01/25/live-like-a-caveman.

Levine, J. A. "Nonexercise Activity Thermogenesis—Liberating the Life-Force." *Journal of Internal Medicine* 262 (2007): 273–87.

Levine, J. A., McCrady, S. K., Boyne, S., Smith, J., Cargill, C., and Forrester, T. "Non-exercise Physical Activity in Agricultural and Urban People." *Urban Studies* 48 (2011): 2417–27.

Liebenberg, L. "Persistence Hunting by Modern Hunter–Gatherers." *Current Anthropology* 47 (2006): 1017–26.

Lieberman, D. E., Bramble, D. M., Raichlen, D. A., and Shea, J. J. "The Evolution of Endurance Running and the Tyranny of Ethnography: A Reply to Pickering And Bunn (2007)." *Journal of Human Evolution* 53 (2007): 439–42.

Lieberman, D. E., Venkadesan, M., Werbel, W. A., Daoud, A. I., D'Andrea, S., Davis, I. S., Ojiambo Mang'Eni, R., and Pitsiladis, Y. "Foot Strike Patterns and Collision Forces in Habitually Barefoot versus Shod Runners." *Nature* 463 (2010): 531–36.

MacArthur, D. "The ACTN3 Sports Gene Test: What Can It Really Tell You?" *Wired Science* (blog), *Wired*, November 30, 2008. http://www.wired.com/wiredscience/2008/11/the-actn3-sports-gene-test-what-can-it-really-tell-you.

MacArthur, D. G., Seto, J. T., Raftery, J. M., Quinlan, K. G., Huttley, G. A., Hook, J. W., Lemckert, F. A., et al. "Loss of ACTN3 Gene Function Alters Mouse Muscle Metabolism and Shows Evidence of Positive Selection in Humans." *Nature Genetics* 39 (2007): 1261–65.

Macera, C. A., Pate, R. R., Powell, K. E., Jackson, K. L., Kendrick, J. S. and Craven, D. E. "Predicting Lower-Extremity Injuries among Habitual Runners." *Archives of Internal Medicine* 149 (1989): 2565–68.

McDougall, C. "Born to Be a Trail Runner." *Well* (blog), *New York Times*, March 18, 2011. http://well.blogs.nytimes.com/2011/03/18/born-to-be-a-trail-runner.

———. *Born to Run: A Hidden Tribe, Superathletes, and the Greatest Race the World Has Never Seen*. New York: Knopf, 2009.

———. "Born to Run Marathons." *Christopher McDougall* (blog), November 6, 2010. http://www.chrismcdougall.com/blog/2010/11/born-to-run-marathons.

———. "The Once and Future Way to Run." *New York Times*, November 2, 2011.

Murphy, J. "What's Your Workout?" *Wall Street Journal*, June 7, 2011. http://online.wsj.com/article/SB10001424052702303745304576357341289831146.html.

Newman, A. A. "Appealing to Runners, Even the Barefoot Brigade." *New York Times*, July 27, 2011. http://www.nytimes.com/2011/07/28/business/media/appealing-to-runners-even-the-shoeless.html.

Noakes, T., and Spedding, M. "Run for Your Life." *Nature* 487 (2012): 295–96.

O'Keefe, J. H., Vogel, R., Lavie, C. J., and Cordain, L. "Achieving Hunter-Gatherer Fitness in the 21st Century: Back to the Future." *American Journal of Medicine* 123 (2010): 1082–86.

———. "Exercise like a Hunter-Gatherer: A Prescription for Organic Physical Fitness." *Progress in Cardiovascular Diseases* 53 (2011): 471–79.

Owen, N., Healy, G. N., Matthews, C. E., and Dunstan, D. W. "Too Much Sitting: The Population Health Science of Sedentary Behavior." *Exercise and Sport Sciences Reviews* 38 (2010): 105–13.

Parker-Pope, T. "Are Barefoot Shoes Really Better?" *Well* (blog), *New York Times*, September 30, 2011. http://well.blogs.nytimes.com/2011/09/30/are-barefoot-shoes-really-better.

Raichlen, D. A., Armstrong, H., and Lieberman, D. E. "Calcaneus Length Determines Running Economy: Implications for Endurance Running Performance in Modern Humans and Neandertals." *Journal of Human Evolution* 60 (2011): 299–308.

Reynolds, G. "Are We Built to Run Barefoot?" *Well* (blog), *New York Times*, June 8, 2011. http://well.blogs.nytimes.com/2011/06/08/are-we-built-to-run-barefoot.

———. "Phys Ed: Is Running Barefoot Better for You?" *Well* (blog), *New York Times*, October 21, 2009. http://well.blogs.nytimes.com/2009/10/21/phys-ed-is-running-barefoot-better-for-you.

Ruiz, J. R., Gomez-Gallego, F., Santiago, C., Gonzalez-Freire, M., Verde, Z., Foster, C., and Lucia, A. "Is There an Optimum Endurance Polygenic Profile?" *Journal of Physiology* 587 (2009): 1527–34.

"Running USA: Running Defies the Great Recession." LetsRun.com, June 16, 2010. http://www.letsrun.com/2010/recessionproofrunning0617.php.

Ruxton, G. D., and Wilkinson, D. M. "Thermoregulation and Endurance Running in Extinct Hominins: Wheeler's Models Revisited." *Journal of Human Evolution* 61 (2011): 169–75.

Sisson, M. *The Primal Blueprint*. Malibu, CA: Primal Nutrition, 2009.

Stanfield, M. "Barefoot Running: Crazy Trend or Timeless Wisdom?" *O&P Edge*, April 2010. http://www.oandp.com/articles/2010-04_06.asp.

Steudel-Numbers, K. L., and Wall-Scheffler, C. M. "Optimal Running Speed and the Evolution of Hominin Hunting Strategies." *Journal of Human Evolution* 56 (2009): 355–60.

Taleb, N. N. *The Black Swan: The Impact of the Highly Improbable*, 2nd ed. New York: Random House Trade Paperbacks, 2010.

van Mechelen, W. "Running Injuries: A Review of the Epidemiological Literature." *Sports Medicine* 14 (1992): 320–35.

Yaeger, S. "Your Body's Biggest Enemy." *Women'sHealth*, November 2009 (last modified June 14, 2010). http://www.womenshealthmag.com/health/sedentary-lifestyle-hazards.

Zorpette, G. "Louis Liebenberg: Call of the Wild." *IEEE Spectrum*, February 2006. http://spectrum.ieee.org/computing/networks/louis-liebenberg-call-of-the-wild.

Chapter 7: Paleofantasy Love

Bird, R. "Cooperation and Conflict: The Behavioral Ecology of the Sexual Division of Labor." *Evolutionary Anthropology* 8(2) (1999): 65–75.

Codding, B. F., Bird, R. B., and Bird, D. W. "Provisioning Offspring and Others: Risk-Energy Trade-Offs and Gender Differences in Hunter-Gatherer Foraging Strategies." *Proceedings of the Royal Society B* 278 (2011): 2502–09.

Darwin, C. *The Descent of Man and Selection in Relation to Sex*. London: John Murray, 1871. Reprinted in *The Origin of Species by Means of Natural Selection; or The Preservation of Favored Races in the Struggle for Life and The Descent of Man and Selection in Relation to Sex*. New York: Modern Library, 1936.

De Waal, F. B. M. *Bonobo: The Forgotten Ape*. Berkeley: University of California, 1998.

———. "Was 'Ardi' a Liberal?" *The Blog* (blog), *Huffington Post*, October 18, 2009. http://www.huffingtonpost.com/frans-de-waal/was-ardi-perhaps-liberal_b_325201.html.

Fortunato, L. "Reconstructing the History of Marriage Strategies in Indo-European–Speaking Societies: Monogamy and Polygyny." *Human Biology* 83 (2011): 87–105.

Geary, D. C. *Male, Female: The Evolution of Human Sex Differences*. 2nd ed. Washington, DC: American Psychological Association, 2010.

Gurven, M., and Hill, K. "Why Do Men Hunt? A Reevaluation of "Man The Hunter" and the Sexual Division of Labor." *Current Anthropology* 50 (2009): 51–74.

Hager, L. D. "Sex and Gender in Paleoanthropology." In *Women in Human Evolution*, edited by L. D. Hager, 1–28. London: Routledge, 1997.

Hammer, M. F., Woerner, A. E., Mendez, F. L., Watkins, J. C., Cox, M. P., and Wall, J. D. "The Ratio of Human X Chromosome to Autosome Diversity Is Positively Correlated with Genetic Distance from Genes." *Nature Genetics* 42 (2010): 830–31.

Hawkes, K., and Bird, R. B. "Showing Off, Handicap Signaling, and the Evolution of Men's Work." *Evolutionary Anthropology* 11 (2002): 58–67.

Hawkes, K., O'Connell, J. F., and Coxworth, J. E. "Family Provisioning Is Not the Only Reason Men Hunt: A Comment on Gurven and Hill." *Current Anthropology* 51 (2010): 259–64.

Holden, C., and Mace, R. "Sexual Dimorphism in Stature and Women's Work: A Phylogenetic Cross-Cultural Analysis." *American Journal of Physical Anthropology* 110 (1999): 27–45.

Hrdy, S. B. *Mother Nature: A History of Mothers, Infants, and Natural Selection*. New York: Pantheon, 1999.

———. *Mothers and Others: The Evolutionary Origins of Mutual Understanding*. Cambridge, MA: Belknap Press of Harvard University Press, 2011.

Ingoldsby, B. B. "Marital Structure." In *Families in Global and Multicultural Perspective*, 2nd ed., edited by B. B. Ingoldsby and S. D. Smith, 99–112. Thousand Oaks, CA: Sage, 2006.

Lovejoy, O. J. "Reexamining Human Origins in Light of *Ardipithecus ramidus*." *Science* 326 (2009): 74e1–8.

McLean, C. Y., Reno, P. L., Pollen, A. A., Bassan A. I., Capellini, T. D., Guenther, C., Indjeian, V. B, et al. "Human-Specific Loss of Regulatory DNA and the Evolution of Human-Specific Traits." *Nature* 471 (2011): 216–19.

Ryan, C. "Sex, Evolution, and the Case of the Missing Polygamists: Were Our Ancestors Polygamists, Monogamists, or Happy Sluts?" *Sex at Dawn* (blog), *Psychology Today*, October 1, 2010. http://www.psychologytoday.com/blog/sex-dawn/201010/sex-evolution-and-the-case-the-missing-polygamists.

Ryan, C., and Jethá, C. *Sex at Dawn: How We Mate, Why We Stray, and What It Means for Modern Relationships*. New York: Harper Perennial, 2010.

Shalit, W. "Is Infidelity Natural? Ask the Apes." *CNN Opinion*, September 2, 2010. http://www.cnn.com/2010/OPINION/09/02/shalit.infidelity/index.html?hpt=C2.

Shultz, S., Opie, C., and Atkinson, Q. D. "Stepwise Evolution of Stable Sociality in Primates." *Nature* 479 (2011): 219–24.

Silk, J. B. "The Path to Sociality." *Nature* 479 (2011): 182–83.

Small, M. F. *What's Love Got to Do with It? The Evolution of Human Mating*. New York: Anchor Books, 1995.

Soulsbury, C. D. "Genetic Patterns of Paternity and Testes Size in Mammals." *PLoS ONE* 5 (2010): e9581.

Stanford, C. B. "The Social Behavior of Chimpanzees and Bonobos." *Current Anthropology* 39 (1998): 399–420.

Trivers, R. L. "Parental Investment and Sexual Selection." In *Sexual Selection and the Descent of Man, 1871–1971*, edited by B. Campbell, 136–79. Chicago: Aldine, 1972.

Zihlman, A. "The Paleolithic Glass Ceiling: Women in Human Evolution." In *Women in Human Evolution*, edited by L. D. Hager, 91–113. London: Routledge, 1997.

Zinjanthropus. "The Sexuality Wars, Featuring Apes." *A Primate of Modern Aspect* (blog), September 6, 2010. http://zinjanthropus.wordpress.com/2010/09/06/the-sexuality-wars-featuring-apes.

Zuk, M. *Sexual Selections: What We Can and Can't Learn about Sex from Animals*. Berkeley: University of California Press, 2002.

Chapter 8: The Paleofantasy Family

Blurton-Jones, N. G., and Marlowe, F. W. "Selection for Delayed Maturity: Does It Take 20 Years to Learn to Hunt and Gather?" *Human Nature* 13 (2002): 199–238.

Bogin, B. "Evolutionary Hypotheses for Human Childhood." *Yearbook of Physical Anthropology* 40 (1997): 63–89.

Cant, M. A., and Johnstone, R. A. "Reproductive Conflict and the Separation of Reproductive Generations in Humans." *Proceedings of the National Academy of Sciences of the USA* 105 (2008): 5332–36.

DeSilva, J. M. "A Shift toward Birthing Relatively Large Infants Early in Human Evolution." *Proceedings of the National Academy of Sciences of the USA* 108 (2011): 1022–27.

Gettler, L. T. "Direct Male Care and Hominin Evolution: Why Male–Child Interaction Is More Than a Nice Social Idea." *American Anthropologist* 112 (2010): 7–21.

Gettler, L. T., McDade, T. W., Feranilc, A. B., and Kuzawa, C. W. "Longitudinal Evidence That Fatherhood Decreases Testosterone in Human Males." *Proceedings of the National Academy of Sciences of the USA* 108 (2011): 16194–99.

Gettler, L. T., and McKenna, J. J. "Evolutionary Perspectives on Mother–Infant Sleep Proximity and Breastfeeding in a Laboratory Setting." *American Journal of Physical Anthropology* 144 (2011): 454–62.

Gibbons, A. "The Birth of Childhood." *Science* 322 (2008): 1040–43.

Gray, P. B. "The Descent of a Man's Testosterone." *Proceedings of the National Academy of Sciences of the USA* 108 (2011): 16141–42.

Hagen, E. H., and Barrett, H. C. "Cooperative Breeding and Adolescent Siblings: Evidence for the Ecological Constraints Model?" *Current Anthropology* 50 (2009): 727–37.

Hrdy, S. B. *The Langurs of Abu: Female and Male Strategies of Reproduction*. Cambridge, MA: Harvard University Press, 1980.

———. *Mother Nature: A History of Mothers, Infants, and Natural Selection*. New York: Pantheon, 1999.

———. *Mothers and Others: The Evolutionary Origins of Mutual Understanding*. Cambridge, MA: Belknap Press of Harvard University Press, 2011.

Kaptijn, R., Thomese, F., van Tilburg, T. G., and Liefbroer, A. C. "How Grandparents Matter: Support for the Cooperative Breeding Hypothesis in a Contemporary Dutch Population." *Human Nature* 21 (2010): 393–405.

Koenig, W. D., and Dickinson, J. L., eds. *Ecology and Evolution of Cooperative Breeding in Birds*. Cambridge: Cambridge University Press, 2004.

Kramer, K. L. "Cooperative Breeding and Its Significance to the Demographic Success of Humans." *Annual Review of Anthropology* 39 (2010): 417–36.

———. "The Evolution of Human Parental Care and Recruitment of Juvenile Help." *Trends in Ecology and Evolution* 26 (2011): 533–40.

Kruger, A. C., and Konner, M. "Who Responds to Crying? Maternal Care and Allocare among the !Kung." *Human Nature* 21 (2010): 309–29.

Mace, R., and Sear, R. "Are Humans Cooperative Breeders?" In *Grandmotherhood: The Evolutionary Significance of the Second Half of Female Life*, edited by E. Voland, A. Chasiotis, and W. Schiefenhoevel, 143–59. Piscataway, NJ: Rutgers University Press, 2005.

Mansbach, A. *Go the F**k to Sleep*. New York: Akashic Books, 2011.

McKenna, J. J., Ball, H. L., and Gettler, L. T. "Mother-Infant Cosleeping, Breastfeeding and Sudden Infant Death Syndrome: What Biological Anthropology Has Discovered about Normal Infant Sleep and Pediatric Sleep Medicine." *Yearbook of Physical Anthropology* 50 (2007): 133–61.

Small, M. F. "Mother's Little Helpers." *New Scientist*, December 7, 2002.

———. *Our Babies, Ourselves: How Biology and Culture Shape the Way We Parent*. New York: Doubleday, 1998.

Strassmann, B. I. "Cooperation and Competition in a Cliff-Dwelling People." *Proceedings of the National Academy of Sciences of the USA* 108 (2011): 10894–901.

Strassmann, B. I., and Gillespie, B. "Life-History Theory, Fertility and Reproductive Success in Humans." *Proceedings of the Royal Society B* 269 (2002): 553–62.

Winking, J., and Gurven, M. "The Total Cost of Father Desertion." *American Journal of Human Biology* 23 (2011): 755–63.

Chapter 9: Paleofantasy, in Sickness and in Health

Barnes, I., Duda, A., Pybus, O. G., and Thomas, M. G. "Ancient Urbanization Predicts Genetic Resistance to Tuberculosis." *Evolution* 65 (2010): 842–48.

Baron, B., and Schembri-Wismayer, P. "Using the Distribution of the CCR5-Δ32 Allele in Third-Generation Maltese Citizens to Disprove the Black Death Hypothesis." *International Journal of Immunogenetics* 38 (2010): 139–43.

Coghlan, A. "Briefing: Cancer Is Not a Disease of the Modern World." *New Scientist*, October 14, 2010.

Crespi, B. J. "The Emergence of Human-Evolutionary Medical Genomics." *Evolutionary Applications* 4 (2011): 292–314.

David, A. R., and Zimmerman, M. R. "Cancer: An Old Disease, a New Disease or Something In Between? *Nature Reviews Cancer* 10 (2011): 728–33.

de Silva, E., and Stumpf, M. P. H. "HIV and the CCR5-Δ32 Resistance Allele." *FEMS Microbiology Letters* 241 (2004): 1–12.

Dickerson, J. E., Zhu, A., Robertson, D. L., and Hentges, K. E. "Defining the Role of Essential Genes in Human Disease." *PLoS ONE* 6 (2011): e27368.

Domazet-Lošo, T., and Tautz, D. "An Ancient Evolutionary Origin of Genes Associated with Human Genetic Diseases." *Molecular Biology and Evolution* 25 (2008): 2699–707.

Donoghue, H. D. "Insights Gained from Palaeomicrobiology into Ancient and Modern Tuberculosis." *Clinical Microbiology and Infection* 17 (2011): 821–29.

Finch, C. E. "Evolution of the Human Lifespan and Diseases of Aging: Roles of Infection, Inflammation, and Nutrition." *Proceedings of the National Academy of Sciences of the USA* 107 (2010): 1718–24.

Gage, T. B. "Are Modern Environments Really Bad for Us?: Revisiting the Demographic and Epidemiologic Transitions." *Yearbook of Physical Anthropology* 48 (2005): 96–117.

Galvani, A. P., and Slatkin, M. "Evaluating Plague and Smallpox as Historical Selective Pressures for the CCR5-Δ32 HIV-Resistance Allele." *Proceedings of the National Academy of Sciences of the USA* 100 (2003): 15276–79.

———. "Intense Selection in an Age-Structured Population." *Proceedings of the Royal Society B* 271 (2004): 171–76.

Glass, W. G., McDermott, D. H., Lim, J. K., Lekhong, S., Yu, S. F., Franks, W. A., Pape, J., Cheshier, R. C., and Murphy, P. M. "CCR5 Deficiency Increases Risk of Symptomatic West Nile Virus Infection." *Journal of Experimental Medicine* 203 (2006): 35–40.

Greaves, M. "Darwinian Medicine: A Case for Cancer." *Nature Reviews Cancer* 7 (2007): 213–21.

Mennerat, A., Nilsen, F., Ebert, D., and Skorping, A. "Intensive Farming: Evolutionary Implications for Parasites and Pathogens." *Evolutionary Biology* 37 (2010): 59–67.

Nerlich, A. G., and Bachmeier, B. E. "Paleopathology of Malignant Tumours Supports the Concept of Human Vulnerability to Cancer." *Nature Reviews Cancer* 7 (2007): 563.

Novembre, J., Galvani, A. P., and Slatkin, M. "The Geographic Spread of the CCR5 Δ32 HIV-Resistance Allele." *PLoS Biology* 3 (2005): e339.

Starr, B. "Is There a Genetic Reason Some People Survived the Plague during the Middle Ages?" Tech Museum, May 12, 2004. http://www.thetech.org/genetics/ask.php?id=10.

Stephens, J. C., Reich, D. E., Goldstein, D. B., Shin, H. D., Smith, M. W., Carrington, M., Winkler, C., et al. "Dating the Origin of the CCR5-Δ32 AIDS-Resistance Allele by the Coalescence of Haplotypes." *American Journal of Human Genetics* 62 (1998): 1507–15.

Waldron, T. "What Was the Prevalence of Malignant Disease in the Past?" *International Journal of Osteoarchaeology* 6 (1996): 463–70.

Wells, S. *Pandora's Seed: The Unforeseen Cost of Civilization*. New York: Random House, 2010.

Chapter 10: Are We Still Evolving? A Tale of Genes, Altitude, and Earwax

Alenderfer, M. S. "Moving Up in the World." *American Scientist* 91 (2003): 542–49.

Allen, J. S. "Why Humans Are Crazy for Crispy." *Chronicle Review*, May 27, 2012.

Andrews, T. M., Kalinowski, S. T., and Leonard, M. J. "'Are Humans Evolving?' A Classroom Discussion to Change Student Misconceptions Regarding Natural Selection." *Evolution Education and Outreach* 4 (2011): 456–66.

Balter, M. "Are Humans Still Evolving?" *Science* 309 (2005): 234–37.

Brantingham, P. J., Rhode, D., and Madsen, D. B. "Archaeology Augments Tibet's Genetic History." *Science* 329 (2010): 1466–67.

Byars, S. G., Ewbank, D., Govindarajuc, D. R., and Stearns, S. C. "Natural Selection in a Contemporary Human Population." *Proceedings of the National Academy of Sciences of the USA* 107 (2010): 1787–92.

Cauchi, S. "Long and Short of It—We're Taller." The Age, April 12, 2004. http://www.theage.com.au/articles/2004/04/11/1081621836499.html#.

Cochran, G., and Harpending, H. *The 10,000 Year Explosion: How Civilization Accelerated Human Evolution*. New York: Basic Books, 2009.

Coyne, J. A. "Are Humans Still Evolving? A Radio 4 Show." *Why Evolution Is True* (blog), August 17, 2011. http://whyevolutionistrue.wordpress.com/2011/08/17/are-humans-still-evolving-a-radio-4-show.

———. "Are We Still Evolving? Part 2." *Why Evolution Is True* (blog), September 18, 2010. http://whyevolutionistrue.wordpress.com/2010/09/18/are-we-still-evolving-part-2.

———. *Why Evolution Is True*. New York: Penguin, 2010.

Darwin, C. *The Descent of Man and Selection in Relation to Sex*. London: John Murray, 1871. Reprinted in *The Origin of Species by Means of Natural Selection; or The Preservation of Favored Races in the Struggle for Life and The Descent of Man and Selection in Relation to Sex*. New York: Modern Library, 1936.

Gibbons, A. "Tracing Evolution's Recent Fingerprints." *Science* 329 (2010): 740–42.

Graves, G. "Feel as Happy as a Pig in Mud!" *Self*, April 2012.

Greg, W. R. "On the Failure of 'Natural Selection' in the Case of Man." *Fraser's Magazine*, September 1868.

Hawks, J. "Human Evolution Stopping? Wrong, Wrong, Wrong." *John Hawks Weblog* (blog), October 10, 2008. http://johnhawks.net/taxonomy/term/304.

Hernandez, R. D., Kelley, J. L., Elyashiv, E., Melton, S. C., Auton, A., McVean, G., 1000 Genomes Project, Sella, G., and Przeworski, M. "Classic Selective Sweeps Were Rare in Recent Human Evolution." *Science* 331 (2011): 920–24.

McKie, R. "Is Human Evolution Finally Over?" *Guardian*, February 2, 2002. http://www.guardian.co.uk/science/2002/feb/03/genetics.research.

Milot, E., Mayer, F. M., Nussey, D. H., Boisverta, M., Pelletierc, F., and Réale, D. "Evidence for Evolution in Response to Natural Selection in a Contemporary Human Population." *Proceedings of the National Academy of Sciences of the USA* 108 (2011): 17040–45.

Pritchard, J. K., and Di Rienzo, A. "Adaptation—Not by Sweeps Alone." *Nature Reviews Genetics* 11 (2010): 665–67.

Pritchard, J. K., Pickrell, J. K., and Coop, G. "The Genetics of Human Adaptation: Hard Sweeps, Soft Sweeps, and Polygenic Adaptation." *Current Biology* 20 (2010): R208–15.

Shepherd, J. A. "Lawson Tait—Disciple of Charles Darwin." *British Medical Journal* 284 (1982): 1386–87.

Simonson, T. S., Yang, Y., Huff, C. D., Yun, H., Qin, G., Witherspoon, D. J., Bai, Z., et al. "Genetic Evidence for High-Altitude Adaptation in Tibet." *Science* 329 (2010): 72–75.

Stearns, S. C., Byars, S. G., Govindarajuc, D. R., and Ewbank, D. "Measuring Selection in Contemporary Human Populations." *Nature Reviews Genetics* 11 (2010): 611–22.

Stock, J. T. "Are Humans Still Evolving?" *European Molecular Biology Organization Reports* 9 (2008): S51–54.

Storz, J. T. "Genes for High Altitudes." *Science* 329 (2010): 40–42.

Tait, L. "Has the Law of Natural Selection by Survival of the Fittest Failed in the Case of Man?" *Dublin Quarterly Journal of Medical Science* 47 (1869): 102–13.

Wade, N. "Adventures in Very Recent Evolution." *New York Times*, July 19, 2010.

———. "Scientists Cite Fastest Case of Human Evolution." *New York Times*, July 1, 2010.

Ward, P. "What Will Become of *Homo sapiens*?" *Scientific American*, January 2009.

Williamson, S. H., Hubisz, M. J., Clark, A. G., Payseur, B. A., Bustamante, C. D., and Nielsen, R. "Localizing Recent Adaptive Evolution in the Human Genome." *PLoS Genetics* 3 (2007): e90.

Yi, X., Liang, Y., Huerta-Sanchez, E., Jin, X., Cuo, Z. X. P., Pool, J. E., Xu, X., et al. "Response to Brantingham et al. 2010." *Science* 329 (2010): 1467–68.

———. "Sequencing of 50 Human Exomes Reveals Adaptation to High Altitude." *Science* 329 (2010): 75–78.

Yoshiura, K., Kinoshita, A., Ishida, T., Ninokata, A., Ishikawa, T., Kaname, T., Bannai, M., et al. "A SNP in the *ABCC11* Gene Is the Determinant of Human Earwax Type." *Nature Genetics* 38 (2006): 324–30.

Index

Page numbers in *italics* refer to illustrations.